Freezing & Drying

Created and designed by the editorial staff of Ortho Books

PROJECT EDITORS

Anne Coolman
Sally W. Smith

WRITER

Charlotte Walker

DESIGNERS

Linda Hinrichs
Karen Berndt

PHOTOGRAPHER

Michael Lamotte

PHOTOGRAPHIC STYLIST

Sara Slavin

ASSOCIATE EDITOR

Beverley DeWitt

Writer/food consultant Charlotte Walker packs home-dried fruits and vegetables.

Ortho Books

Publisher
Robert L. Iacopi

Editorial Director
Min S. Yee

Managing Editors
Anne Coolman
Michael D. Smith

System Manager
Mark Zielinski

Senior Editor
Sally W. Smith

Editors
Jim Beley
Diane Snow
Deni Stein

System Assistant
William F. Yusavage

Production Manager
Laurie Sheldon

Photographers
Laurie A. Black
Michael D. McKinley

Photo Editors
Anne Dickson-Pederson
Pam Peirce

Production Editor
Alice E. Mace

Production Assistant
Darcie S. Furlan

National Sales Manager
Garry P. Wellman

Operations/Distribution
William T. Pletcher

Operations Assistant
Donna M. White

Administrative Assistant
Georgiann Wright

Address all inquiries to:
Ortho Books
Chevron Chemical Company
Consumer Products Division
575 Market Street
San Francisco, CA 94105

Copyright © 1984
Chevron Chemical Company
Printed in the United States of America
All rights reserved under international
and Pan-American copyright
conventions.

First Printing in March, 1984

1 2 3 4 5 6 7 8 9
84 85 86 87 88 89

ISBN 0-89721-027-1
Library of Congress Catalog Card
Number 83-62653

Acknowledgments

Food Stylists
Kathy Briggs
Rose Hansen
Amy Nathan

Design
Carol Kramer
Cathy McAuliffe
Phil Offenhauser

Illustrator
Carol Kramer

Additional Photography
Page 67: Tom Tracy

Research and Text
Ann Halverson

Recipe Consultants
Patricia Litman
Ann Ny Pang
Jaime Sheridan

Recipe Tester
Toby Walker

Typography
Turner & Brown, Inc.
Forestville, CA

Color Separations
Color Tech Corp.
Redwood City, CA

Copyediting
Editcetera
Berkeley, CA

We also thank the following
individuals and organizations for their
contributions to this book:
Bee Beyer
Bill Cagle
Creative Sports
Kathy Hadley
Marmot Mountain Works
Sears
Dr. George York
Paulette De Jong
Cooperative Extension Service
University of California

Front Cover
The just-picked goodness of fresh
fruits, vegetables, and herbs is yours
to enjoy the year around when you
freeze or dry them at their peak.

Back Cover
Upper left: Freezing fresh fruits,
page 17
Upper right: Freezing baked goods,
page 47
Lower left: Recipes for snack mixes,
page 77
Lower right: Freezing asparagus,
page 28

Chevron Chemical Company
575 Market Street, San Francisco, CA 94105

Freezing & Drying

Preface

Preserving seasonal food is not only smart, it's economical. Frozen at home during the height of their season, fresh foods will retain their peak flavor and texture for months.

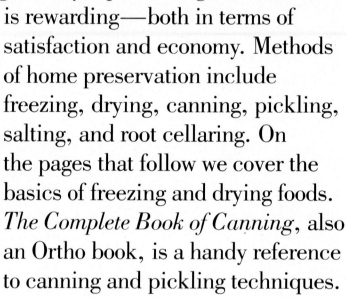

Stocking your cupboards and freezer with home-preserved foods is well worth the effort. Whether you're a home gardener faced with baskets of produce at its peak or simply want to take advantage of seasonal good buys, preserving foods at home is rewarding—both in terms of satisfaction and economy. Methods of home preservation include freezing, drying, canning, pickling, salting, and root cellaring. On the pages that follow we cover the basics of freezing and drying foods. *The Complete Book of Canning*, also an Ortho book, is a handy reference to canning and pickling techniques.

In many homes, the freezer is the center of an informal food storage system. Freezing is easy; fast; and maintains food color, flavor, and nutrients. It is also suited to more foods than any other method of preservation.

Drying is one of the oldest forms of food preservation. Dried foods have an advantage over frozen in that they require no expensive equipment to keep them at their best—just a shelf in a cool, dark, dry place. However, while freezing maintains more of the original fresh flavor and texture of foods, the drying process alters their texture significantly. Compare, for example, frozen and dried apple slices. Frozen apples are plump, juicy, and can be used in much the same way as their fresh counterparts—in a pie or cobbler. Dried apple slices are flat and chewy. They have a concentrated flavor and are best when reconstituted for fruit compotes or eaten out-of-hand as a sweet snack.

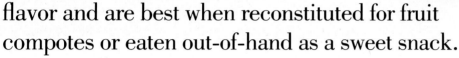

This book is a comprehensive guide to freezing and drying all kinds of foods using the best up-to-date methods and recipes available. It will be an invaluable guide for the novice who is just deciding which preservation method to use, as well as for the experienced cook looking for a quick reference.

The following five chapters cover techniques and recipes for freezing everything from peaches to pies. The last chapter is a detailed guide to drying foods, flavorings, and fragrant or decorative accents for your home. A special section on smoke-cooking foods in a covered barbecue or water smoker concludes the book.

Packaging protects frozen-food quality. Airtight rigid plastic containers are excellent.

Freezing Fundamentals

Freezing has many advantages over other methods of food preservation. Frozen food tastes fresher and it also retains its original color, texture, flavor, and nutritional value better than food preserved by other methods. And freezing is one of the simplest ways to "put up" food for later enjoyment.

Freezer storage allows you to take advantage of special sales, stocking up when prices are low or when favorite or seasonal produce is plentiful. You can save up to half of what you'd spend for green beans, for example, in the winter by home-freezing them when they're inexpensive and at the peak of freshness.

This chapter will familiarize you with the essentials of freezing. After reading it, you'll be able to freeze everything from the summer squash growing in your garden to a robust chili for a crowd. The chapter also serves as a handy general reference for the next four chapters, which outline freezing techniques for specific foods.

The home freezer has revolutionized meal preparation. It provides more free time for the cook, makes advance preparation feasible, expedites last-minute meals for unexpected guests, and is a money-saver.

Heavy-duty aluminum foil molds to foods of various sizes and shapes, providing a tight wrap.

Basic Rules for Successful Freezing

By observing the following basic rules, you can successfully preserve the flavor, color, texture, and nutritive qualities of almost any food.

Select high-quality foods. Freezing retains quality, but can't improve it.

Use suitable packaging materials to protect frozen food against air and loss of moisture. See pages 11 and 12.

Freeze foods immediately after packing.

Limit the amount of food to be frozen within a 24-hour period to no more than 3 pounds of food per cubic foot of freezer storage space.

Store food at 0°F or below.

Observe recommended storage periods. See the charts in the chapters that follow.

Plan a constant turnover of the freezer's contents so that no food is kept too long. First in, first out is a good rule of thumb.

Thaw frozen foods carefully.

Home Freezing

The only exposure many households have to home freezing is the freezer section of their refrigerator. Although the refrigerator's freezer compartment is terrific for quickly consumed, in-and-out foods like ice cream, it is not the best choice for long-term storage. Because the refrigerator-freezer tends to be opened frequently, its temperature fluctuates. Its size also limits the quantity of food that can be stored at any given time. And, if frozen food is to maintain its quality over a long period of time, it must be stored at 0°F or lower, a temperature it's difficult for some refrigerator-freezers to reach and for many to maintain. For each 10° above zero, the storage life of your food is cut up to half.

Freezer compartments inside single-door refrigerators are the least effective and are not suited for long-term storage; every time the refrigerator door is opened, the freezer is subject to temperature fluctuations. Two-door refrigerator-freezers come in three configurations: with the freezer on the top, bottom, or side. Side-by-side combinations offer more freezer space than top or bottom models—but proportionately less refrigerator space. Purchase price and operating costs may be higher than for the other two configurations. Narrow shelves can also be inconvenient for bulky items. Bottom-mounted freezers put fresh foods at eye level, for convenience, but mean that you must stoop to retrieve frozen items. Some experts believe that top-mounted freezers offer the best dollar value per cubic foot of storage space.

If you're purchasing a new refrigerator-freezer, be sure to select one with dual temperature controls. With a single temperature control, it may be impossible to maintain your freezer at 0°F without also freezing the contents of your refrigerator. Other purchase considerations are similar to those for selecting an upright or a chest freezer. Refer to consumer guides for specific recommendations on brands and models.

Is a Home Freezer for You?

A home freezer could be one of the wisest investments you'll ever make. Compared with other forms of preserving, freezing's only major disadvantage is the purchase price of the freezer. But the advantages freezing offers—maintenance of food quality, simplicity of food preparation, and ability to preserve a great variety of foods—outweigh this disadvantage in many people's opinion.

You may want to give freezing a try in a small way before investing in a home freezer. If you can maintain a temperature of about 0°F in your two-door refrigerator-freezer, follow the instructions in this book to freeze some of your favorite foods. Since your freezer temperature may fluctuate, storage times will not be as long as those cited in the charts, but with limited freezer space you will use your frozen food more quickly.

If you want to purchase and freeze larger quantities of meat or produce, freezer rental space can also act as an introduction. It is relatively inconvenient, however (you'll need to plan your visits and transfer one or two weeks' worth of food at a time), and it is not available everywhere. Check the Yellow Pages of your telephone directory for space in your area.

Purchasing a Home Freezer

Every year, manufacturers introduce new freezer models offering improved features. How do you determine the freezer type, size, and features that will suit your needs? The following information should help. Also, consult manufacturers' handbooks for detailed descriptions of the models you are interested in. Check the warranties as well.

FREEZER TYPES

Chest. Chest freezers open from the top. They are the less expensive of the two basic freezer types to operate because less cold air escapes each time you open them. (The more air that escapes, the harder a freezer must work to maintain a constant cold temperature.) Chest freezers offer more usable space than uprights, but they also take up more floor space, and unless storage is well planned, packaged foods toward the bottom of the chest can be difficult to locate and reach. Many chest models now come with sliding or lift-out baskets to make it easier to find and retrieve foods.

Upright. Upright freezers open like a conventional refrigerator and require less floor space than chest models. Interior shelves, door shelves, and bins make it easy to locate stored food. Small leftovers and frozen juice concentrates, as well as odds and ends (coffee beans, bread crumbs), can be conveniently stored and found in the door shelves. The major drawback to uprights is their higher operating cost due to loss of cold air each time the door is opened.

MANUAL OR AUTOMATIC DEFROST?

Manual defrosting is a tedious and messy job that must take place every six months, or when the frost layer inside the freezer builds up to ¼ to ½ inch thick. An automatic-defrost system continuously and automatically removes the frost. Generally, automatic-defrost freezers consume more energy and therefore are more expensive to operate than manual-defrost models. However, a buildup of frost in a manually operated system ultimately causes the freezer to work harder, and frost buildup reduces storage capacity.

FREEZER CAPACITY

One of the most important freezer selection considerations is capacity. Freezers are available in sizes as small as 5 cubic feet and as large as 27 cubic feet. As a general guideline, allow 5 to 6 cubic feet per family member. A family of four typically requires a freezer with a capacity of 20 to 24 cubic feet. Freezers operate most efficiently when they are at least three-quarters full. A 20-cubic-foot freezer

barely half full wastes purchase cost, floor space, and the energy required to cool the empty space.

CONSTRUCTION

A freezer's construction is a major key to its ability to maintain low temperatures and prevent air loss. Adequate insulation keeps foods frozen up to 36 hours should your freezer lose power. Look for 3 to 4 inches of insulating material, except in the newer thin-wall-insulation models. All-steel construction extends the life of the freezer. A tight-fitting door with "cushion" gaskets all around prevents air loss.

When you're shopping for a home freezer, look for the yellow energy label attached to each freezer model in the store; it tells you the estimated operating cost of the model. Even if the price is somewhat higher for a model with a lower operating cost, lower utility bills can quickly pay back the extra money spent.

WHAT NOT TO FREEZE

Just about any food can be frozen, but some foods maintain their fresh taste and texture better than others. When you are uncertain about whether to freeze a particular food, freeze a sample. After several days, thaw the sample and note its appearance and taste. Although the test does not indicate how well the food will hold up in storage over time, it does give you an idea about the effect freezing has on it.

The following list indicates foods that are *not* recommended for the freezer and explains why:

Salad greens, cucumbers, radishes, green peppers, tomatoes, and green onions lose their crispness and become limp.

Potatoes become mushy or mealy and may darken.

Canned hams can become watery and tough.

Stuffed poultry cannot be frozen safely because stuffing in the center of the bird does not freeze quickly enough to prevent the growth of bacteria.

Most fried foods lose their crispness.

Half-and-half, sour cream, and cottage cheese separate and become grainy. Buttermilk and yogurt do the same, but you can use them in baking.

Mayonnaise and egg- or cream-based salad dressings separate.

Cooked egg white becomes tough and rubbery.

Boiled or fluffy frostings made with egg whites turn sticky.

Meringues toughen.

Most gelatin dishes weep.

The flavor of cloves, garlic, fresh onion, and synthetic vanilla becomes stronger when these foods are frozen.

It's a good idea to monitor the temperature of your refrigerator's freezer section with a freezer thermometer, available at hardware and kitchen-equipment stores. Place the thermometer toward the top and front of the freezer. Leave it overnight without opening the freezer door. If the temperature is above 0°F the next day, adjust your freezer's temperature control and take another reading.

WHERE TO PUT A FREEZER

When deciding where to place your new freezer, look for a space that is both large enough and convenient—a kitchen or pantry is usually ideal. A garage, although perhaps convenient and spacious, is often a poor location because of temperature variations. Summer heat will cause the freezer to work harder to maintain a constant temperature, adding to energy costs. Avoid placing the freezer in a draft or near a source of hot air, such as an oven, a clothes dryer, or a furnace duct.

The air around the freezer should be as dry as possible. Moisture can rust the outside and cause frost buildup inside. Leave space behind, around, and above the freezer—3 to 4 inches above and 3 inches on all sides—for air circulation and easy cleaning. Allow space for the door to open fully.

Get acquainted with your freezer by reading its use and care book. It contains important information on how to operate your particular model. And remember that periodic cleaning and good maintenance will extend the life of your freezer.

BASIC FREEZER MANAGEMENT

A freezer in which foods are arranged in a logical manner—frozen produce in one part, meat in another—will hold more packages than one that is disorganized. Divide your freezer into sections and designate particular areas: Vegetables, Fruits, Meats, Casseroles and Main Dishes, Baked Goods, and Desserts. Use a basket, box, or divider to keep little items together and make them easier to locate. Rotate packages so that the oldest are at the front or top of the freezer where you will see and use them first.

THE FREEZER INVENTORY

It's a good idea to keep a list of the frozen foods you have in storage. Update the list each time you put foods in or take them out. A running inventory helps you plan menus, keep a balanced assortment of foods on hand, and remember to use foods before they have outlived their storage life. Design a sheet that lists frozen foods by category. List the number of packages and the freezing date. Each time you remove or add a package, make a note of it. Cross off items as you deplete the supply. Here's an example.

Food	No. of Packages	Date Frozen	No. Pkgs. Removed
Fruits			
Raspberries	14 pts.	6/19/83	III
Peach slices	12 qts.	8/6/83	THL II
Apple slices	8 qts.	10/3/83	II

STORAGE TIMES

The length of time frozen food can be stored depends on the care with which the food was handled before freezing, the type of food, the quality of the packaging materials, and maintenance of the proper storage temperature (0°F or less). Recommended storage periods, which have been determined by thorough research and experimentation, are listed for specific foods in the chapters that follow. The storage tables in this book assume the use of a freezer that maintains a temperature of 0°F or less.

Keep frozen produce only from one growing season to the next; other foods should be stored for shorter periods. For

---

FREEZER OPERATION: SAVING ENERGY AND MONEY

These tips will help you reduce the cost of operating your freezer.

Place your freezer in a cool, dry area where the temperature is consistent.

A full freezer operates more efficiently than one that is almost empty.

Open the freezer door as rarely as possible. Try to remove food for one meal at one time.

If your model has a "power saver," use it according to the manufacturer's directions. Generally, the heater should be turned off when the weather is dry and on when the weather is humid.

Check the door seals occasionally to be sure they're clean and that they seal properly. Replace deteriorated seals.

When adding food to the freezer, limit the amount to what will freeze within 24 hours. A general rule of thumb is to add no more than 3 pounds of food per cubic foot of freezer capacity at one time.

Defrost manual freezers regularly.

Keep the condenser coils clean.

example, meats like bacon and seasoned sausage should be kept no longer than a month because the salt they contain accelerates the development of rancidity in the fat. Store ground and sliced meats and cut-up poultry for only about a month, because their exposed surfaces make them more readily subject to oxidation.

PACKAGING FOODS FOR FREEZING

Packaging and wrapping materials must be moistureproof and vaporproof (resistant to dampness and air), strong enough to withstand handling at low temperatures, tasteless, and capable of being tightly closed or sealed.

There are two types of packaging materials for home-freezing use—rigid containers and flexible bags or wrappers.

RIGID CONTAINERS

Rigid containers are recommended for foods that are soft or liquid at room temperature. Containers include glass freezer jars, ovenproof casseroles, plastic freezer containers, aluminum foil containers, wax-coated cardboard cartons, tin cans, and ice cube trays or muffin tins. Rigid containers are often reusable; therefore, although their initial purchase cost may be higher than that of wrapping materials, recycling the containers saves money over time.

When purchasing rigid containers, consider the number of servings you want to freeze together. Freeze meal-size amounts to avoid leftovers.

Glass freezer jars (which are the same as canning jars) are especially designed with extra-wide mouths and tapered sides to permit removal of food without complete thawing. Two-piece screw-on caps with built-in rubber rings make an airtight, leakproof seal. Jars are available in ½-pint, pint, and 1½-pint sizes.

Ovenproof casseroles and dishes can be used for freezing, heating, and serving. Seal covers with freezer tape. Slip those without lids into plastic bags and twist the top of the bag tight. Or tape on a cover of heavy-duty aluminum foil. Some ovenproof casseroles can go directly from the freezer to a preheated oven. If you do not want to tie up your casseroles and dishes, wrap the food in casserole-wrap fashion. (See page 13.)

Plastic freezer containers stack easily, are reusable, and are available in many sizes and shapes. Choose containers with tight-fitting lids.

Aluminum foil containers are convenient and relatively inexpensive. Foods can go from freezer to oven to table in the container. (However, foil containers cannot be used in most microwave ovens.)

Properly handled, most fruits and vegetables retain their fresh taste, texture, and color when frozen.

Packaging and wrapping materials suitable for freezing include, from top to bottom: (1) rigid plastic containers, glass freezer jars, and ovenproof casseroles; (2) aluminum foil containers, tin cans, and wax-coated cardboard cartons; (3) heavy-duty aluminum foil, plastic food storage bags, coated freezer paper, and freezer tape.

Wax-coated cardboard cartons come with plastic liners or bags. They are the least expensive containers to use and are available in ½-pint, pint, quart, and 2-quart sizes.

Tin cans—for example, shortening and coffee cans—are excellent for packaging foods that are delicate or easily mashed. Line these cans with a plastic food-storage bag and seal the tops with freezer tape since they are not airtight.

Ice cube trays and muffin tins make handy containers for small amounts. Tray-freeze (see page 14) foods until firm; then remove and pack in plastic freezer bags.

FLEXIBLE BAGS AND FREEZER WRAPS

Flexible bags and freezer wraps include plastic food-storage bags designed for freezing, boil-in bags, aluminum foil, clear plastic wrap designed for freezing, and coated or laminated freezer papers. Most meats, baked goods, bulky vegetables, and fruit packed without sugar or syrup can be packaged successfully in bags or in freezer wrap.

Plastic food-storage bags designed for freezing are available in pint, quart, 2-quart, and 1-gallon sizes. (Some plastic food-storage bags on the market are not suitable for freezing. Generally, the heavier the bag, the better it will perform. Check the label on the package; if it does not say that the bags are suitable for freezing, choose another bag.)

Boil-in bags, a recent development in frozen food packaging, are made of a material that withstands temperatures from below 0° to 240°F. After filling, the bags are heat-sealed using a sealing machine. Cooked foods and blanched vegetables, sealed and frozen in these bags, can be heated to serving temperature simply by dropping the bags in boiling water or placing them in a microwave oven.

Aluminum foil is an excellent and readily available wrapping material for freezing. It is strong and flexible at low temperatures and can be folded and pressed to make tight seams and squeeze out all air. Use either heavy-duty foil or a double thickness of regular foil.

Plastic wrap designed for freezing is effectively moistureproof, vaporproof, and resistant to tearing, and it clings closely to food. (Like plastic food-storage bags, some plastic wraps on the market are not suitable for freezing. Check the label on the package to make sure.)

Freezer paper is used primarily to wrap meats. It is coated on one side to make it vaporproof and moistureproof. Place the food on the coated (shiny) side of the paper.

Freezer tape is especially designed for sealing packages for the freezer. The adhesive remains effective at very low temperatures and the tape won't let go when it becomes cold or wet, as other tapes often do. Freezer tape not only makes an airtight seal, it also gives a surface on which to label packages. If your supermarket doesn't stock freezer tape, try a hardware store or a catalog-sales store.

Waxed paper and uncoated butcher paper provide *inadequate* protection for frozen foods. Rewrap food in these wrappings with suitable freezer wrapping material.

PACKAGING TIPS

These packaging tips ensure fresher tasting, high-quality frozen foods.

Allow for headspace. Foods expand during freezing. If you do not allow space for expansion, packages will bulge, lids will pop off, bags can split, and glass jars can shatter. For dry packs, such as berries packed without syrup, allow ½-inch headspace. Liquid and semiliquid packs, such as pasta sauce, must have ½-inch headspace in pint containers with wide-mouth openings; 1 inch in quarts. Ideally, leave just enough space so that the food expands flush with the top of the container, leaving no air pockets.

Remove air. Air pockets between the food and the packaging material collect moisture drawn from the food, resulting in frost and freezer burn. To exclude air, fill jars and containers completely, allowing only for the required headspace. Run the blade of a knife around the inside of the container to eliminate air pockets. When wrapping food, press out as much air as possible and mold the wrapping as close to the food as possible. To remove air

Wrapping Techniques

There are four techniques for wrapping foods for freezing. Choose the wrapping material and technique depending on the type of food you are wrapping.

DRUGSTORE WRAP—for flat or rectangular-shaped foods such as steaks or chops: Cut off enough wrapping material to wrap around the food about 1½ times. Place the food in the center of the wrap and bring opposite edges of the wrap together over the top so that they meet. Repeatedly crease and fold the edges together until the fold lies against the food. Press the fold down across the food, squeezing out the air. Fold one end of the packaging material in; then fold that end once and bring it up against the food. Repeat with the other end. Seal with freezer tape.

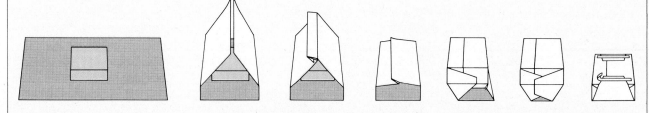

BUTCHER WRAP—for bulky items such as roasts: Use enough wrapping material to wrap around the food twice. Place the food at one corner of the wrap and turn it and the paper over at the same time. Tuck the side edges up around the food and roll the food toward the opposite corner. Bring up the end and seal with freezer tape.

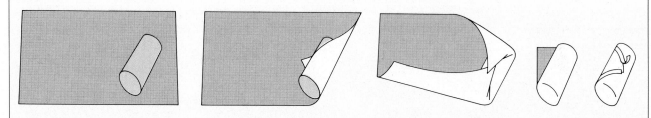

BUNDLE WRAP—for odd-shaped foods and foods that are slightly moist: Place food in the center of a square of heavy-duty aluminum foil large enough to wrap it completely. Bring the four corners together in a pyramid shape. Squeeze and mold the foil close to the food. Fold ends over and press against the package. Seal with freezer tape.

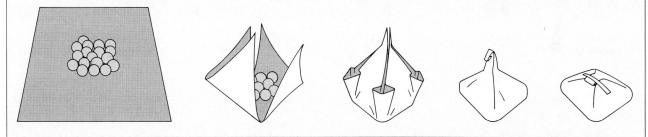

CASSEROLE WRAP—Permits recycling of containers after contents are frozen: Line casserole with heavy-duty aluminum foil, leaving 1½-inch collar around the edges. Place food in casserole; cover with a piece of foil the size of casserole and collar. To seal, press out air from center toward sides; fold edges over; press together. Freeze. When frozen firm, remove food package from casserole, label, and return to freezer.

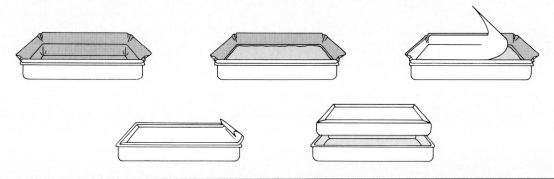

In tray-freezing, un-wrapped foods—like these blanched vegetable pieces—are frozen on a baking sheet until firm. When packaged together for storage, the individual pieces remain separate and thaw quickly.

from plastic bags, insert a drinking straw, inhale to draw out the air, and fasten tightly. Or use the water-dip method: Immerse the bag in a sinkful of water, taking care that no water enters the bag. This pushes the plastic against the food, forcing out all air. Then twist the top of the bag tight and seal.

Seal containers. Containers designed for freezer use have tight-fitting lids. Lids that are not airtight should be sealed with freezer tape; wrap tape around the outer edge of the lid where it meets the container. Or place the container inside a freezer bag and twist the top of the bag tight. Containers that do not have lids should be sealed with heavy-duty aluminum foil and freezer tape.

Labels enable you to identify foods in the freezer, their age, and the amount in the package. Mark each package with the kind of food, date of storage, recipe used (if any), weight or number of servings, type of pack, variety of fruit or vegetable, and intended use (if relevant). Freezer tape, a pen with permanent ink, and self-adhesive labels are handy for labeling. To make it easy to identify foods in the freezer, you may want to color-code each food category—blue labels for vegetables, yellow for meats, and so on.

FREEZING AND THAWING POINTERS

HOW TO FREEZE

Freeze foods only after they have cooled to room temperature. In each 24-hour period, freeze a maximum of 3 pounds of food per cubic foot of freezer space. Overloading slows down the freezing process, and foods frozen too slowly lose quality and may spoil. To preserve freshness and flavor, follow the freezing guidelines for specific foods in the chapters that follow.

Quick-freezing is the recommended method of freezing packaged foods. To quick-freeze, place packages in a single layer close to the outside freezer walls, where freezing plates or coils are located. Allow space between packages so that air can circulate freely around them. Freeze packages 24 hours before sorting and stacking them. Refer to your freezer manual—some new models are equipped with a quick-freeze section or shelf where foods freeze more rapidly.

Tray-freezing is an excellent method for freezing "piece" foods like meatballs, peas and beans, hors d'oeuvres, drop cookies, and fruits that can be packed without syrup. One advantage of this method is that pieces freeze individually and remain separate, allowing you to remove only as many as you need. Thawing is quicker, too. Another reason for tray-freezing is to firm up foods like cakes and pies before packaging so that the packaging material will not adhere to them. To tray-freeze, spread unwrapped food on a baking sheet and freeze just until firm. Then immediately package as usual.

Maintaining Quality During Storage

Almost all food, whether plant or animal in origin, is subject to deterioriation and eventual decay. To minimize loss of quality—and, in severe cases, contamination—in frozen foods, you need to control the growth of microorganisms and enzymes and to avoid oxidation.

Unlike canning, freezing does not sterilize foods by destroying the microorganisms present in them. Cold temperatures only inhibit the growth of microorganisms and slow down enzyme activity and oxidation. During thawing, spoilage organisms start to grow again; as the temperature of the food rises, so does the rate of growth of the organisms. Enzymes cause fruits and vegetables to mature and ripen, and eventually to overripen and decay; freezing inhibits enzyme activity. Fruits are often packed in sugar syrup and ascorbic acid to control enzyme activity. Vegetables are blanched to slow down the decay process and help retain flavor, color, and nutritive value. Oxidation causes meat and poultry fat to become rancid. The best protection against oxidation is to store food in airtight packages and follow recommended storage periods.

The following clues warn you that foods may have suffered deterioration during freezing. Proper packaging, temperatures, and storage periods are the primary keys to maintaining quality. If you discover food with any of these warning signs, chances are it is still safe to eat, but the flavor will no longer be of top quality. You don't need to throw the food out unless it has developed off-odors, but plan to use it as soon as possible and take its condition into consideration when deciding on a cooking method. (Use a steak in a stew instead of broiling it, for example.)

Ice buildup. Icy crystals inside the package indicate that the food has thawed, at least partially, and refrozen or that the food didn't freeze quickly. To minimize the size of ice crystals, freeze foods as quickly as possible and maintain them at 0°F or lower.

Freezer burn. When foods are not carefully packaged, moisture is drawn out of them by the dry air of the freezer, resulting in freezer burn: a dry surface with grayish-white spots. To protect against freezer burn, use proper packaging materials and be certain that packages are airtight.

Texture change. The texture of foods that have thawed and refrozen deteriorates. Vegetables become soft and limp, meats toughen slightly, and the pasta in combination dishes may become soggy from absorbing moisture from the sauce.

Color change. When foods are stored in the freezer too long, their color fades. Vegetables, for example, turn a dull, drab color. See the charts in the chapters that follow for optimum storage times.

Sauce consistency. When food has passed its optimum storage period or has been allowed to thaw and refreeze, the consistency of the sauce may change. Starches may break down and sauces may not be smooth.

Thawing Frozen Food

Some foods do not need to be thawed before cooking; others will retain their flavor and texture best if partially or completely thawed. Vegetables, unbaked pies, and fish can be taken directly from the freezer and cooked or baked. Meats and poultry, on the other hand, will retain their juices and flavor better when completely thawed before cooking.

Thawing in the refrigerator is the safest, although slowest, thawing method. When you thaw foods at room temperature, you run the risk of spoilage. If time is short, some foods, particularly poultry and meats, may be placed (in their freezer wrappings) in a plastic bag and thawed in a basin of cold water; this cuts thawing time about in half. A microwave oven is by far the fastest and easiest method of all. Your oven manual lists appropriate settings and approximate defrosting times.

Thaw most foods in their freezer wrappings. An exception is quick breads and cakes, which can be affected by the moisture that collects on the inside of the package as they thaw.

Refreezing Thawed Food

If food feels cold, is still firm, and contains ice crystals, you may refreeze it without cooking it first. Be aware, however, that refreezing involves a loss of food quality and flavor and that refrozen food cannot be kept as long as freshly frozen food.

Once food has thawed, organisms that can cause spoilage begin to multiply. Therefore, unless you cook it first, do not refreeze food that has thawed completely. (You can, for example, cook thawed raw meat and then refreeze it.) Foods that have been held between 32° and 40°F for more than 24 hours should be used as soon as possible; they should not be refrozen. Seafood spoils quickly and should not be refrozen; cook it and use as soon as possible. Casseroles or dishes that have been precooked should not be refrozen. Discard any foods that have developed off-odors or colors.

To refreeze foods, spread the packages out in the freezer so that cold air can circulate around them. Label foods as refrozen, date, and use as soon as possible.

Freeze a variety of seasonal fresh fruits and enjoy the taste of summer the year around.

Freezing Fruits, Vegetables & Herbs

Although no preserving method can maintain all the qualities of fresh food, freezing comes closest to this goal for most fruits, vegetables, and herbs. Because fruits and vegetables do suffer some loss of texture and give up moisture more readily as they thaw, you'll find that frozen produce tastes closest to fresh when you cook or process it in some way before serving. Herbs are best when added to cooked dishes.

This chapter provides step-by-step instructions for preparing, packing, and freezing most fruits and vegetables and shows you how to preserve the just-picked flavor of fresh herbs.

You'll find tantalizing recipes that capture the sun-ripened sweetness of fruits and berries—from freezer jam to a Fresh Fruit Frosty drink that can be whipped up from frozen fruit. Or stock your freezer with make-ahead vegetable dishes like Carrots Grand Marnier and Zucchini Purée; and try the pungent basil-based Pesto Sauce for winter pasta and soups.

The more flavor your produce has before freezing, the more it will retain during storage, so it's essential to freeze only the freshest, most flavorful fruits and vegetables you can obtain.

Wax-coated cardboard cartons are inexpensive and come in a variety of sizes: from tiny ones for single servings to meal-size versions.

For freezing, select homegrown or purchased produce that is fully ripe—preferably vine, bush, or tree ripened—and firm. Underripe produce, although firm, is not as flavorful or as sweet as mature produce. Examine each fruit or vegetable carefully. Choose only those that are fresh, firm, and free of bruised spots, blemishes, or any indications of spoilage.

SELECTING PRODUCE FOR FREEZING

HOMEGROWN PRODUCE

The well-planned garden yields just enough fruits and vegetables to eat fresh in season and the right amount, at the right time, to preserve. Plan your garden so that you will not be faced with bushels of produce all ready to be frozen at the same time. Successive plantings can be made all summer long, provided that plants mature before the first frost. See the charts on pages 21 and 28 to determine harvest seasons.

Save your garden plan from year to year. Record the varieties that did well, and use this information to eliminate mistakes and to maximize garden space.

Pick produce in the morning when it is still cool; it will stay fresh and flavorful longer. Place the picked fruits and vegetables in containers made of cloth, wood, or a loosely woven material, rather than in plastic bags or metal containers. Open containers permit air to circulate and heat to dissipate. Your produce will retain its fresh quality longest if cooled immediately after it is harvested.

To preserve flavor and texture, it's essential to prepare and freeze produce as soon as possible after it's harvested. On warm days, a noticeable loss of sweetness can be detected in peas, asparagus, and sweet corn only a few minutes after harvesting. If you cannot freeze foods at once, refrigerate them immediately after harvesting and keep refrigerated until you are free to freeze them.

PURCHASED PRODUCE

It's not essential to have your own garden or orchard in order to obtain fruits and vegetables suitable for freezing. There are many ways for nongardeners to obtain really fresh produce.

One of the best is to shop the countryside at U-pick farms that sell directly to the public. Many states and counties publish farm trail maps listing farmers who sell produce directly from their farms. You can also check the want ads in your local newspaper. Or, write to the chambers of commerce in nearby small towns and ask where fresh produce is available for picking or purchase. Not only will you save money, but the freshness, flavor, and texture will be much better than that of store-bought produce. Take a cooler with you on your produce-buying excursions. Pack a picnic in it to enjoy along the way, then fill it with the produce you buy, to keep it cool and fresh during the trip home.

You can also purchase freshly harvested fruits and vegetables at farmers' markets in urban areas. Watch for produce trucks parked along highways, too.

Food co-ops often carry very fresh, inexpensive produce. If you don't belong to a co-op, ask the buyer at your local supermarket to special-order an extra quantity of produce for you. Or check your local wholesale produce terminal; if you're willing to purchase in quantity, you may be able to buy at the wholesale price.

Fruits and vegetables are sold by the bushel, the box, the crate, the peck, or the lug. How many pounds of a specific fruit or vegetable each of these containers holds depends on the shape and density of the produce in it. The charts on page 26 indicate the weight of a full container of certain fruits and vegetables and what the yield will be after they are prepared for freezing. The yield assumes that the item has been peeled or shelled, pitted, and cut up or sectioned in the customary manner. If a lug of peaches is too much for you to use, share it with neighbors and friends. When you purchase produce in boxes, remember to check the quality at the bottom of the container.

With a little forethought you can plant a garden especially suited for preserving, with produce ripening throughout the summer.

For freezing, select fruit that is firm, fully ripe, and free of bruised spots or blemishes.

Peel, core, and slice apples to prepare them for freezing. Pack them with sugar, in syrup, or in fruit juice. Homemade applesauce freezes beautifully, too.

HOW TO PREPARE AND PACK FRUITS

With the exception of avocados, fruits can be frozen successfully. Since freezing affects the texture of fruits—they break down somewhat and lose juice during thawing—it affects their shape or form but it doesn't reduce their flavor.

Frozen fruits are at their best when cooked before serving or used in recipes that call for puréeing. Frozen peaches, for example, are excellent baked in pies and cobblers. Whirled in a blender, they become a fruit sauce or a base for blender drinks or homemade ice cream. Since they become slightly mushy when thawed completely, they aren't at their best served as is for dessert, raw in salads or compotes, or as a garnish. To use them this way, serve before they thaw completely.

For specific instructions on preparing and packing particular fruits, see pages 24 and 25. But in general, when you're ready to start freezing, wash the fruit thoroughly; then work quickly to peel, core, and cut it. Avoid long periods of soaking and standing. Some fruits can be frozen whole; others need to be halved or sliced. You can leave the skins on small whole fruits like plums; however, peel fruits like peaches that are halved or

sliced. To peel easily, dip fruit in boiling water for about 30 seconds, plunge into cold water, and slip off the skins. Fruits like peaches and apricots darken when they are peeled and cut. To prevent this, add an antioxidant during packing. See page 22.

CHOOSING A PACKING METHOD

There are three ways to pack fruit for freezing: syrup pack, sweetened dry pack, and unsweetened dry pack. Sugar is the key ingredient in both the syrup pack and the sweetened dry pack. It helps preserve the fruit by reducing enzymatic oxidation, which causes spoilage, and helps maintain texture and shape. Fruits packed by the syrup or the sweetened dry pack method remain plumper and firmer than fruits that are frozen without sugar.

Almost any fruit can be packed in syrup. Fruits with little natural juice, such as apples, are always best frozen in a syrup pack. You can pack fruits in extra-light, light, medium, or heavy syrup concentrations, to suit your taste. The lighter syrups do not mask the fruit's natural flavor and are lower in calories, but heavy syrup maintains the form of the fruit better than the lighter syrups.

Plain water is not recommended as a packing medium because it tends to leach out the fruit's natural flavor, sweetness, and color. For a lower calorie, less sweet product, substitute bottled or canned fruit juice for the sugar syrup. Pineapple, cranberry, and apple juices work well. Without a high concentration of sugar, however, fruit shape and texture may suffer somewhat.

Sweetened dry pack is a good method for juicy fruits like plums, peaches, and apricots. The sugar draws out the natural juices of these fruits and combines with them to form a syrup. (However, this method produces less liquid than the syrup pack.) The California Cooperative Extension Service of the U.S. Department of Agriculture recommends a ratio of 1 part sugar by weight to 4 or 5 parts fruit by weight. The chart on pages 24 and 25 reflects this recommendation. We have found, however, that you can reduce the amount of sugar to suit your taste and to reduce calories. Just be aware that this may compromise shape and texture.

The unsweetened dry pack method is the easiest way to freeze fruit, but because no sugar is used, texture and shape will suffer. If you plan to cook the fruit after thawing or to use it in blender drinks or puréed sauces, this may not be a consideration. Also, without sugar to help retard enzymatic activity, fruits in an unsweetened dry pack have a shorter freezer life than those packed in syrup or a sweetened dry pack: 3 to 6 months rather than 8 to 12 months.

INSTRUCTIONS FOR PACKING

Syrup pack. Select the syrup concentration you prefer from the chart. To prepare, combine sugar and water in a saucepan. Bring to a boil, stirring to dissolve the sugar. Chill before using. You can prepare the syrup a day or two ahead of time and store it in the refrigerator.

SYRUP CONCENTRATIONS

	Sugar*	Water	Yield
Extra-light	1 cup	4 cups	4½ cups
Light	2 cups	4 cups	5 cups
Medium	3 cups	4 cups	5½ cups
Heavy	4¾ cups	4 cups	6½ cups

NOTE: Syrup may be flavored with citrus peel or spices to taste. Remember that spices intensify during freezing.

*Honey or light corn syrup may be substituted for up to about one-fourth of the granulated sugar. Higher proportions will give a very different flavor. Brown sugar and molasses should not be used because their flavor and color will overpower that of the fruit.

To pack fruits, pour ½ cup syrup into an empty freezer container. (To prevent darkening, add an antioxidant to the syrup for certain fruits; see page 22, "How to Protect Fruit Color.") Add the whole, halved, or sliced fruit, pouring in more syrup as needed to cover the fruit. Top fruit with crumpled waxed paper, foil, or plastic wrap to keep it submerged in the syrup and prevent darkening.

Sweetened dry pack. Halve or slice fruit into a bowl or shallow pan. Sprinkle with ascorbic acid if necessary. (See page 22, "How to Protect Fruit Color.") Add the sugar and gently mix with the fruit until juice forms and the sugar dissolves. Package in rigid containers. Or spread on baking sheets in a single layer and tray-freeze. (See page 14.) When the pieces are firm, pack in freezer bags or containers.

Unsweetened dry pack. Place fruit in a single layer on baking sheets. Treat with an antioxidant if necessary. (See page 22, "How to Protect Fruit Color.") Tray-freeze until firm. (See page 14.) Pack the pieces in plastic freezer bags or containers.

CONTAINERS FOR FREEZING FRUITS

Containers that protect the quality of fruits in the freezer include heavily waxed, plastic, or aluminum containers; glass freezer jars; plastic freezer bags; and heat-sealed pouches. Pack fruit and juice or syrup tightly into containers. Leave headspace, following these guidelines.

	Wide-top container		Narrow-top container	
	Pint	Quart	Pint	Quart
Syrup pack	½"	1"	¾"	1½"
Sweetened and unsweetened dry packs	½"	½"	½"	½"

Special tools from your hardware or department store simplify fruit and vegetable preparation. Cherry pitters neatly push seeds from cherries; sawtooth corn cutters remove kernels from ears; vegetable and apple peelers make fast work of this task; melon ballers create uniform spheres; and French bean slicers turn out uniform shreds in minutes.

HARVEST SEASON

Fruit	JANUARY	FEBRUARY	MARCH	APRIL	MAY	JUNE	JULY	AUGUST	SEPTEMBER	OCTOBER	NOVEMBER	DECEMBER
Apples						▓	▓	▓	▓	▓	▓	
Apricots						▓	▓					
Bananas	▓	▓	▓	▓	▓	▓	▓	▓	▓	▓	▓	▓
Blackberries							▓	▓				
Blueberries	▓					▓	▓	▓				
Cherries						▓	▓					
Coconut									▓	▓	▓	
Cranberries									▓	▓	▓	
Currants						▓	▓					
Figs						▓	▓	▓	▓	▓		
Grapefruit	▓	▓	▓	▓								▓
Grapes								▓	▓	▓		
Kiwi fruit	▓	▓	▓	▓	▓	▓					▓	▓
Mangoes					▓	▓	▓	▓				
Muskmelons					▓	▓	▓	▓	▓	▓		
Nectarines						▓	▓	▓				
Oranges	▓	▓	▓	▓							▓	▓
Peaches						▓	▓	▓	▓			
Pears								▓	▓	▓		
Persimmons										▓	▓	▓
Pineapple				▓	▓	▓						
Plums						▓	▓	▓	▓			
Raspberries						▓	▓					
Rhubarb				▓	▓	▓						
Strawberries				▓	▓	▓						
Watermelon						▓	▓	▓				

Freezer Jams and Jellies

Freezer jams and jellies glisten with the natural color and flavor of fresh fruit. They are thinner than traditional cooked jams and jellies, but are simple, fast, and fresh tasting.

To make freezer jams, crush the fruit and then mix it with sugar and commercial fruit pectin (which causes the preserves to set up). Virtually any fleshy fruit, from blueberries to cherries, makes a beautiful freezer jam. (See the recipe for Spicy Blueberry Freezer Jam at right.) Save firmer fruits, such as apples and pears, for conserves or chutneys.

Pack jams and jellies in jars or rigid containers, allow ½-inch headspace (for pints), and seal with tight-fitting lids. Then let the containers stand at room temperature until the jam or jelly sets—up to 24 hours—before freezing. Freezer jams and jellies will keep in the freezer 8 months. Thawed, they can be stored in the refrigerator up to 3 weeks.

Thawing Fruits

Fruits may be thawed at room temperature, in the refrigerator, or under cold running water. Leave containers or freezer bags unopened, turning them several times to ensure uniform thawing. One pint of fruit in syrup will thaw in 4 to 6 hours in the refrigerator, 2 to 3 hours at room temperature, and 30 to 60 minutes under cold running water.

Use thawed fruit as soon as possible because flavor loss and darkening occur rapidly after thawing. Fruits you plan to eat without cooking or puréeing will have the best texture if they are eaten while a few ice crystals remain. Frozen fruit may be cooked without thawing.

How to Protect Fruit Color

Some fruits will brown when exposed to air after peeling or cutting. The chart on pages 24 and 25 tells you which fruits are susceptible to this problem. Preparing and freezing these fruits quickly reduces the amount of discoloration; however, you will also need to protect them with an antioxidant. The most common color protectors are ascorbic acid, commercial antioxidants, and lemon juice.

Ascorbic acid, sold in drugstores, is available in powder and crystalline forms. For syrup packs, dissolve the ascorbic acid in several tablespoons of cold water and add it to the syrup just before you pour it over the fruit. For sweetened dry packs, dissolve the ascorbic acid in cold water and sprinkle it over the fruit before adding the sugar. For unsweetened dry packs, sprinkle the ascorbic acid–water solution over the fruit just before tray-freezing.

Commercial antioxidants—Fruit Fresh and Ever-Fresh are two—are a combination of ascorbic and citric acid and sugar. They are sold in supermarkets. Follow the manufacturers' instructions for correct proportions and dissolving methods.

Lemon juice adds its flavor to the fruit, which you may or may not like. It is also less effective than ascorbic acid or commercial antioxidants. Dip fruit in a solution of 1 tablespoon lemon juice per quart of water.

Spicy Blueberry Freezer Jam

- 2½ pints blueberries
- 1 tablespoon lemon juice
- ½ teaspoon ground cinnamon
- ⅛ teaspoon ground nutmeg
- 5 cups sugar
- ¾ cup water
- 1 package (1¾ oz) powdered fruit pectin

1. In a large bowl, crush blueberries one layer at a time. (Crushing too many berries at a time inhibits the free flow of the juice.)

2. Measure 3 cups of the crushed blueberries and place in a large pan. Stir in lemon juice, cinnamon, and nutmeg. Thoroughly mix sugar into fruit. Let stand 10 minutes.

3. Combine water and fruit pectin in a saucepan. Bring to a full boil and boil 1 minute, stirring constantly.

4. Stir hot pectin liquid into fruit and continue to stir vigorously for 3 minutes to distribute pectin.

5. Ladle into freezer containers, leaving ½-inch headspace for pints. Cover with lids and let stand at room temperature until set. (It may take up to 24 hours.) Store jam in freezer up to 8 months.

Yield: About 6½ cups

STRAWBERRY DESSERT SAUCE

Keep this dessert sauce on hand as an instant topping for ice cream, cake, or crêpes.

2 pints fresh strawberries, hulled

1 cup (10 oz) currant jelly

2 tablespoons brandy

1 tablespoon lemon juice

Combine strawberries, currant jelly, brandy, and lemon juice in a blender or food processor. Process until smooth.

Yield: 4 cups

TO FREEZE: Spoon into small, rigid freezer containers, leaving ¼-inch headspace. Freeze up to 1 year.

FRESH FRUIT FROSTY

Fresh, cool, and easy, this frosty fruit drink is right for any summer day. Try substituting peeled and sliced peaches or nectarines for the strawberries.

1 pint strawberries, hulled and sliced

¼ cup plain yogurt

1 egg

3 tablespoons sugar

2 teaspoons fresh lime juice

Mint and/or whole strawberries for garnish

1. Arrange berries in a shallow pan and freeze until firm. A few minutes before serving, remove from freezer.

2. In a food processor or blender, blend yogurt, egg, sugar, and lime juice. With motor running, drop in frozen berries, a few at a time. Process until smooth. Serve at once or hold in freezer, removing several minutes before serving to soften slightly. Garnish with mint or whole berries if desired.

Yield: 4 servings

PEACH SLUSH

You can also make this quick summer refresher with melon chunks or strawberries.

1 tablespoon lemon juice

1 cup water

4 to 5 peaches

½ cup buttermilk or milk

2 to 3 tablespoons sugar, or more to taste

Pinch ground cinnamon

1. Combine lemon juice and water in a bowl. Peel and pit peaches; cut into 1-inch chunks. (You should have about 2 cups.) Drop into lemon water. Drain and lightly pat dry. Spread on baking sheets and freeze until firm.

2. Place buttermilk and sugar in a blender or food processor. With motor running, add frozen peach chunks a few at a time until mixture turns to slush. Taste for sweetness; add more sugar if desired. Serve in tall glasses with a dusting of cinnamon.

Yield: 2 servings

Crimson cranberries and golden raisins create a freezer relish that complements savory foods.

RAISINBERRY RELISH

This zesty, colorful relish makes a fine accompaniment for ham, game, or the holiday bird.

2¼ cups golden raisins

2 cups orange juice

1 cup water

¼ cup lemon juice

⅔ cup sugar

3 cups fresh or frozen cranberries

1 tablespoon finely grated orange peel

1. In a 3-quart saucepan, combine raisins, orange juice, water, lemon juice, and sugar. Bring to a boil over high heat, stirring to dissolve sugar. Reduce heat and simmer 10 minutes.

2. Add cranberries and simmer 5 minutes. Add orange peel and simmer about 5 minutes more, until liquid barely covers solid ingredients. Cool to room temperature. Store, covered, in refrigerator up to 1 month.

Yield: About 4½ cups

TO FREEZE: Spoon relish into rigid freezer containers and freeze up to 1 year.

GUIDE TO FREEZING FRUITS

This guide explains which packing methods are best for specific fruits and tells you how to prepare each fruit for freezing. Refer to page 20 for an explanation of packing methods, page 22 for information on antioxidants (ascorbic acid, lemon juice), and page 21 for packing instructions. Before selecting a packing method, think about the end use of the fruit you are freezing. If you want to make pies from some of your frozen peaches, for example, you'll probably want to freeze the peach slices in pie-size quantities and use the sweetened dry pack method rather than the syrup pack. Note on the container or bag the amount of sugar you add before freezing, and then subtract that amount from the recipe when you prepare the pie filling. Do this with any fruit that you might later use in a recipe requiring sugar. Syrup-pack and sweetened-dry-pack fruits can be frozen 8 to 12 months; unsweetened-dry-pack fruits, 3 to 6 months.

Apples
SWEETENED DRY PACK: Peel, core, and slice fruit into a solution of 3 tablespoons lemon juice to 1 gallon cold water. After 2 or 3 minutes, transfer slices to baking sheets lined with paper towels and drain thoroughly. Remove towels and add 1 part sugar to 4 parts apples; mix well. Pack in containers.

SYRUP PACK: Peel, core, and slice fruit. Pack in containers in syrup, adding ½ teaspoon ascorbic acid to each quart of syrup.

APPLESAUCE: Prepare as usual; cool. Pack in containers.

BAKED APPLES: Prepare and bake as usual; cool. Freeze until firm on baking sheets. Pack in containers or, for convenience, store individually in small containers or plastic freezer bags. Thaw and serve cold or reheat.

Apricots
Peel; halve or quarter; remove pits.

SWEETENED DRY PACK: Add ¼ teaspoon ascorbic acid and ½ cup sugar to each quart of fruit; mix well. Pack in containers.

SYRUP PACK: Pack in containers in syrup, adding ¾ teaspoon ascorbic acid to each quart of syrup.

CRUSHED: Coarsely crush fruit. Add ¼ teaspoon ascorbic acid and 1 cup sugar to each quart of crushed fruit; mix well. Pack in containers.

Bananas
Peel and mash ripe fruit. Add 1 to 1½ teaspoons lemon juice to each pint of mashed fruit. Pack in containers. Use in baked breads and cakes or in blender drinks.

Berries
(see also "Cranberries"; "Strawberries")
UNSWEETENED DRY PACK: Spread in single layers on baking sheets; tray-freeze until firm. Transfer to containers or plastic freezer bags.

SWEETENED DRY PACK: Add ¼ to ½ cup sugar to each quart of berries, depending on sweetness; mix well. Pack in containers.

Cherries
Sour: Remove stems; pit if desired.
SWEETENED DRY PACK: Add ¾ to 1 cup sugar to each quart of fruit; mix well. Pack in containers.

SYRUP PACK: Pack in containers in heavy syrup.

CRUSHED: Coarsely crush pitted cherries. Add 1 to 1½ cups sugar to each quart of crushed fruit; mix well. Pack in containers.

Sweet:
UNSWEETENED DRY PACK: Leave stems on. Spread in single layers on baking sheets; tray-freeze until firm. Transfer to containers or plastic freezer bags.

SYRUP PACK: Remove stems; pit if desired. Pack in containers in syrup, adding ½ teaspoon ascorbic acid to each quart of fruit.

Coconut
Split fruit in half; freeze milk for other uses. Remove fruit and grate or grind.

UNSWEETENED DRY PACK: Pack in containers.

SWEETENED DRY PACK: Add ½ cup sugar to each 6 cups grated coconut; mix well. Pack in containers.

Cranberries
UNSWEETENED DRY PACK: Spread in single layers on baking sheets; tray-freeze until firm. Transfer to containers or plastic freezer bags. Overwrap sealed bags of purchased cranberries.

SAUCE: Prepare as usual; cool. Pack in containers.

Currants
UNSWEETENED DRY PACK: Spread in single layers on baking sheets; tray-freeze until firm. Transfer to containers or plastic freezer bags.

SWEETENED DRY PACK: Add ¾ to 1 cup sugar to each quart of fruit; mix well. Pack in containers.

Figs
Remove stems, and peel if desired. Pack whole, halved, or sliced.

UNSWEETENED DRY PACK: Spread in single layers on baking sheets; tray-freeze until firm. Transfer to containers or plastic freezer bags.

SYRUP PACK: Pack in containers in syrup, adding ¾ teaspoon ascorbic acid to each quart of syrup.

Grapefruit
Peel, cutting deep enough to remove white membrane under skin. Section, removing membrane and seeds.

UNSWEETENED DRY PACK: Spread in single layers on baking sheets; tray-freeze until firm. Transfer to containers or plastic freezer bags.

SYRUP PACK: Pack in containers in syrup. Liquid may be part juice from fruit.

Grapes

Remove stems and, if desired, halve and seed varieties with seeds.

UNSWEETENED DRY PACK: Spread whole fruit in single layers on baking sheets; tray-freeze until firm. Transfer to containers or plastic freezer bags.

SYRUP PACK: Pack in containers in light syrup.

Kiwi fruit

Peel. Cut off ends; slice.

SWEETENED DRY PACK: Add 1 cup sugar to each quart of fruit; mix well. Pack in containers.

SYRUP PACK: Pack in containers in syrup.

Mangoes

Peel and slice, avoiding stringy flesh near the seed.

SWEETENED DRY PACK: Add ½ cup sugar to each 5 to 6 cups mango slices; mix gently. Allow to stand until sugar is dissolved.

SYRUP PACK: Pack in containers in syrup.

Melon

Halve, remove seeds, and peel. Cut into slices, cubes, or balls.

UNSWEETENED DRY PACK: Spread in single layers on baking sheets; tray-freeze until firm. Transfer to containers or plastic freezer bags. Use within 1 month.

SWEETENED DRY PACK (for cantaloupe or watermelon): Add 1 pound sugar to each 5 pounds of fruit; mix well. Pack in containers.

SYRUP PACK: Pack in containers in light syrup. For flavor add 1 teaspoon lemon juice to each cup of syrup.

JUICE PACK: Pack fruit in pineapple or orange juice or ginger ale.

Nectarines

Peel; halve or slice, removing pits.

SYRUP PACK: Pack in containers in syrup, adding ½ teaspoon ascorbic acid to each quart syrup.

Oranges

Freeze any variety except navels, which become bitter when frozen. Peel, cutting deep enough to remove white membrane under skin. Section, removing membrane and seeds.

SYRUP PACK: Pack in containers in syrup. Liquid may be part juice from fruit.

Peaches

Peel; halve or slice, removing pits.

SWEETENED DRY PACK: Add ½ teaspoon ascorbic acid and ⅔ cup sugar to each quart of fruit; mix well. Pack in containers.

SYRUP PACK: Pack in containers in syrup, adding ½ teaspoon ascorbic acid to each quart of syrup.

PURÉE: Mash fruit; press through strainer or whirl in blender or food processor. Add 1 cup sugar and ½ teaspoon ascorbic acid to each quart of purée; mix well. Pack in containers.

Pears

Peel and core. Quarter or slice.

SYRUP PACK: Pack in containers in syrup, adding ¾ teaspoon ascorbic acid to each quart of syrup. (Note: Syrup packing is the only method that retains the shape, texture, and color of pears; do not pack by any other method.)

Persimmons

UNSWEETENED DRY PACK: Spread whole fruit on baking sheets; tray-freeze until firm. Transfer to containers or plastic freezer bags. Use within 3 months.

PURÉE: Mash fruit; press through wire strainer or whirl in blender or food processor. Add 1 cup sugar to each 6 cups purée; mix well. Pack in containers.

Pineapple

Peel; remove core and eyes; cut in wedges, cubes, sticks, or thin slices or crush.

SYRUP PACK: Pack in containers in light syrup.

Plums
(including fresh prune plums)

Halve or quarter and pit.

UNSWEETENED DRY PACK: Spread whole fruit on baking sheets; freeze solid. Transfer to containers or plastic freezer bags. Use within 3 months.

SWEETENED DRY PACK: Add 2¼ cups sugar to each 4 pounds prepared plums; mix well. Pack in containers.

SYRUP PACK: Pack in containers in syrup, adding 1 teaspoon ascorbic acid to each quart of syrup.

Rhubarb

Wash, trim, and cut into 1- to 2-inch pieces. Water-blanch (see page 28) for 1 minute; cool immediately in cold water. Drain.

UNSWEETENED DRY PACK: Spread in single layers on baking sheets; tray-freeze until firm. Transfer to containers or plastic freezer bags.

SYRUP PACK: Pack in containers in syrup.

Strawberries

Hull fruit.

UNSWEETENED DRY PACK: Spread in single layers on baking sheets; tray-freeze until firm. Transfer to containers or plastic freezer bags.

SWEETENED DRY PACK: Slice berries or crush slightly if desired. Add ¾ to 1 cup sugar to each quart of fruit; mix well. Pack in containers.

SYRUP PACK: Pack whole or sliced berries in containers in syrup.

Tomatoes

See "Guide to Freezing Vegetables," pages 30 and 31.

FRUIT: PURCHASE WEIGHT AND YIELD		
Fruit	**Fresh, as purchased or picked**	**Frozen**
Apples	"Northwest" Box (40 pounds)	27 to 32 pints
	1¼ to 1½ pounds	1 pint
Apricots	Lug (24 pounds)	30 to 36 pints
	⅔ to 4/5 pound	1 pint
Berries (but see strawberries)	12-basket tray, ½-pint baskets (6 to 8 pounds)	4 to 6 pints
	2 to 3, ½-pint baskets	1 pint
Cantaloupe	1 dozen (28 pounds)	22 pints
	1 to 1¼ pounds	1 pint
Cherries	Lug (23 to 27 pounds)	15 to 18 pints
	1¼ to 1½ pounds	1 pint
Coconut	1 to 1¼ coconuts	1 pint
Cranberries	1 box (25 pounds)	50 pints
	1 peck (8 pounds)	16 pints
	½ pound	1 pint
Currants	2 quarts (3 pounds)	4 pints
	¾ pound	1 pint
Figs	5 to 6 pound box	6 to 7 pints
	¾ to 1 pound	1 pint
Grapefruit	2 medium (2 pounds)	1 pint
Grapes	Lug (28 pounds)	14 to 16 pints
	4 basket crate (20 pounds)	10 to 12 pints
	2 pounds	1 pint
Kiwi fruit	6 to 8 kiwi fruit	1 pint
Mangoes	2 to 3 medium mangoes	1 pint
Oranges	3 to 4 medium oranges	1 pint
Peaches; Nectarines	Bushel (46 to 50 pounds)	30 to 50 pints
	Lug (average 20 pounds)	13 to 20 pints
	1 to 1½ pounds	1 pint
Pears	Lug (24 to 28 pounds)	20 to 25 pints
	Pear box (46 pounds)	37 to 46 pints
	1 to 1¼ pears	1 pint
Persimmons	Lug (24 to 28 pounds)	20 to 25 pints
	2 to 3 medium	1 pint
Pineapple	5 pounds	4 pints
Plums	Lug (average 25 pounds)	16 to 25 pints
	3 quarts (5 pounds)	4 to 5 pints
	1 to 1½ pounds	1 pint
Rhubarb	Lug (25 to 30 pounds)	25 to 45 pints
	⅔ to 1 pound	1 pint
Strawberries	12 basket tray, pint baskets (12 to 14 pounds)	9 to 10 pints
	1⅓ pints	1 pint

VEGETABLES: PURCHASE WEIGHT AND YIELD		
Vegetable	**Fresh, as purchased or picked**	**Frozen**
Artichokes	10 to 12 tiny artichokes 1¼ inches when trimmed	1 pint
Asparagus	1 crate (28 to 35 pounds)	18 to 25 pints
	1 to 1½ pounds	1 pint
Beans, lima, in pods	1 bushel (32 pounds)	12 to 16 pints
	2 to 2½ pounds	1 pint
Beans, snap	1 crate (30 pounds)	30 to 45 pints
	⅔ to 1 pound	1 pint
Beets, without tops	1 bushel (52 pounds)	35 to 42 pints
	1 lug (30 to 32 pounds)	20 to 26 pints
	1¼ pounds	1 pint
Broccoli	1 crate (20 pounds)	20 pints
	1 pound	1 pint
Brussels sprouts	4 quart boxes	6 pints
	1 pound	1 pint
Cabbage	1 to 1½ pounds	1 pint
Carrots, without tops	1 crate (50 pounds)	32 to 40 pints
	1¼ to 1½ pounds	1 pint
Cauliflower	2 medium heads	3 pints
	1⅓ pounds	1 pint
Celery	1 pound	1 pint
Chayote	1½ to 2 pounds	1 pint
Corn, in husks	3 dozen crate	6 to 9 pints
	5 dozen crate	10 to 12 pints
	2 to 2½ pounds (6 to 8 ears)	1 pint
Eggplant	1 to 1½ pounds	1 pint
Okra	1 to 1½ pounds	1 pint
Parsnips	1¼ to 1½ pounds	1 pint
Peas	1 crate (30 pounds)	12 to 18 pints
	2 to 3 pounds	1 pint
Pumpkin	1½ pounds	1 pint
Spinach	1 to 1½ pounds	1 pint
Squash, summer	1 bushel (40 pounds)	32 to 40 pints
	1 lug (25 to 28 pounds)	20 to 28 pints
	1 to 1¼ pounds	1 pint
Squash, winter	1½ pounds	1 pint
Sweet potatoes	⅔ pound	1 pint

Source: Cooperative Extension, U.S. Department of Agriculture, University of California at Davis

Since asparagus is available for such a short time in the spring, freeze it for later enjoyment.

How to Prepare and Pack Vegetables

Most vegetables that are cooked before you eat them are naturals for freezing. In fact, despite the slight loss of texture that occurs during freezing, most frozen vegetables taste very much like fresh after cooking.

Vegetables are at their flavorful best when picked and frozen young. Overmature vegetables will be tough and flavorless after freezing. For preparation instructions for specific vegetables, see the chart on pages 30 and 31. But in general, when you're ready to freeze, work quickly. Wash and sort vegetables by size to ensure even blanching. Prepare only the quantity of vegetables that you can handle in the amount of time available. Keep extra produce refrigerated.

Choosing a Blanching Method

Almost all vegetables need to be blanched and then cooled quickly before freezing.

Blanching slows down the enzyme activity that causes vegetables to ripen and eventually to deteriorate. It also heightens color and makes vegetables tender and pliable so that you can fit more in a freezer container. If they are not blanched before freezing, vegetables develop off-flavors, become discolored and tough, and have a shorter freezer storage life. Blanching can be done in boiling water, over steam, or in a microwave oven.

Almost any vegetable can be blanched successfully in boiling water and most can be blanched over steam. (Do not steam-blanch vegetables that must be treated with an antioxidant to protect color; see the chart on pages 30 and 31.) Steaming protects the shape of vegetables better and also conserves more nutrients than water-blanching, but it takes a bit longer. Microwaving—the newest way of blanching vegetables—is quick, easy, and protects both nutrients and color.

The chart on pages 30 and 31 gives blanching times for individual vegetables. At altitudes above 5,000 feet, add 1 minute to water-blanching and 2 minutes to steam-blanching times.

Instructions for Blanching

Water-blanching. You'll need a large pot with a tight-fitting lid, ideally one equipped with a strainer for lifting the vegetables in and out of the boiling water. If your pot doesn't have its own strainer, use a basket, colander, or wire-mesh strainer that is large enough to hold at least 1 pound of vegetables and that can be easily lifted in and out of the pot. Avoid iron and copper pots.

Allowing 1 gallon of water per pound of vegetable, bring water to a hard boil. Lower vegetable-filled strainer or basket into boiling water; cover pot and return water to boiling point as quickly as possible. Start counting blanching time as soon as water returns to a boil. It's essential to adhere to the recommended blanching time—underblanching results in a loss of color and nutrients during freezing; overblanching affects flavor and texture. Properly blanched vegetables are firm yet tender and are heated through to the center. You may blanch several batches of the same type of vegetable without changing the water.

Harvest Season

Vegetable	January	February	March	April	May	June	July	August	September	October	November	December
Artichokes			X	X								
Asparagus			X	X								
Beans						X	X	X				
Beets						X	X	X	X	X		
Broccoli	X	X	X	X	X	X	X	X	X	X	X	X
Brussels sprouts	X	X	X	X	X	X	X	X	X	X	X	X
Cabbage	X	X	X	X	X	X	X	X	X	X	X	X
Carrots	X	X	X	X	X	X	X	X	X	X	X	X
Cauliflower	X	X	X	X	X	X	X	X	X	X	X	X
Celery	X	X	X	X	X	X	X	X	X	X	X	X
Corn					X	X	X	X	X			
Eggplant							X	X	X			
Mushrooms	X	X	X	X	X	X	X	X	X	X	X	X
Okra							X	X	X			
Onions	X											
Parsnips		X	X									
Peas			X	X	X	X	X					
Peppers							X	X				
Potatoes	X	X	X	X	X	X	X	X	X	X	X	X
Pumpkin									X	X	X	
Rutabagas	X	X	X									
Spinach			X	X	X							
Summer squash						X	X	X	X			
Tomatoes						X	X	X	X			
Turnips	X	X	X									
Winter squash									X	X	X	X

When time is up, immediately plunge blanched vegetables into ice water or hold under cold running water until completely cooled. Then drain vegetables well on absorbent toweling to prevent large ice crystals from forming during freezing.

Steam-blanching. You'll need a large pot with a tight-fitting lid and a steaming rack that fits the pot. Fill the pot with about 2 inches of water and place the rack in the pot. The water should not touch the rack. Bring water to a boil, loosely pack vegetables on rack in a single layer no more than 2 inches deep, and cover the pot. After steaming the recommended length of time, cool and drain vegetables as directed for water-blanching.

Microwave-blanching. You'll need appropriate ovenproof pans or dishes. Follow your oven manufacturer's manual for exact blanching instructions.

CONTAINERS FOR FREEZING VEGETABLES

The most convenient packaging materials for vegetables are plastic freezer bags, rigid containers, heat-sealed pouches, and wax-coated cartons with plastic bags or liners. The coated cartons give vegetables a uniform shape, make stacking easier, and protect the plastic liner from tears. Heat-sealed pouches simplify cooking and cleanup; just drop the pouch directly into boiling water.

Package and freeze vegetables in meal-size quantities. Or tray-freeze (see page 14) and store in plastic freezer bags or other containers so that you can remove only the amount you need for a meal, a single serving, or a soup addition.

Package puréed vegetables like sweet potatoes and pumpkin in rigid containers. Allow ½- to 1-inch headspace.

Wrap corn on the cob in aluminum foil or other freezer wrapping material. Small amounts of green pepper or onion can also be wrapped in foil.

COOKING FROZEN VEGETABLES

Most vegetables should be cooked while still frozen. Exceptions are greens like spinach and Swiss chard, which can be partially thawed to make separation easier and to speed cooking. Corn on the cob should be completely thawed before cooking. Because blanching and freezing tenderize vegetables, cooking time will be half to two-thirds that of fresh.

MARINARA SAUCE

Turn vine-ripened tomatoes into this well-seasoned tomato sauce.

- 3 tablespoons olive oil
- 1 large onion, chopped
- 2 cloves garlic, minced
- 4 pounds very ripe tomatoes, peeled, seeded, and puréed
- 2 teaspoons finely chopped fresh oregano or 1 teaspoon crumbled dried oregano
- 2 teaspoons finely chopped fresh basil or 1 teaspoon crumbled dried basil
  Pinch of sugar
  Salt and pepper

1. Heat oil in heavy skillet. Sauté onion until limp and transparent. Add garlic and sauté 30 seconds.

2. Stir in tomato purée and seasonings. Cook over medium-high heat, stirring frequently, until liquid is reduced and sauce thickens. Adjust seasonings to taste.

Yield: About 2 pints

TO FREEZE: Cool sauce to room temperature. Spoon into rigid freezer containers, leaving ½-inch headspace for pints. Freeze up to 4 months.

Pick or purchase red-ripe tomatoes during the summer and make several batches of Marinara Sauce. Freeze it for year-round use on pasta and pizza or in savory dishes. Jars of the sauce make welcome gifts for family and friends.

Carrots Grand Marnier

3 tablespoons butter
*1 pound carrots, cut in
    1-inch pieces*
    Pinch of thyme
    Salt and pepper
*2 tablespoons Grand
    Marnier*

1. Melt butter in a skillet. Add carrots and sprinkle with thyme and salt and pepper. Turn carrots to coat with butter. Cover pan tightly and cook over low heat for about 15 minutes.

2. Add Grand Marnier; cover and continue cooking until almost tender-crisp.

Yield: 4 servings

To freeze: Cool carrots to room temperature. Freeze in heat-sealed plastic food pouches or rigid freezer containers up to 6 months.

Creamy Cauliflower Purée

*1 large head cauliflower,
    trimmed and broken
    into flowerets*
*1 medium onion,
    coarsely chopped*
*1/4 cup buttermilk or
    half-and-half*
*3 tablespoons butter or
    margarine, softened*
    *Salt, black pepper, and
    pinch cayenne*
*1 to 2 tablespoons butter
    or margarine*
    Chopped parsley

1. In a medium saucepan, cook cauliflower and onion in just enough water to cover for 15 minutes, or until cauliflower is tender when pierced. Drain.

2. In a food processor or blender, whirl cauliflower, onion, buttermilk, and the 3 tablespoons butter until smooth and creamy.

3. Season to taste with salt, black pepper, and cayenne.

4. Spoon mixture into individual (approximately 6-oz) buttered ramekins or ovenproof custard cups. Let cool, wrap tightly with heavy-duty aluminum foil, and freeze. Purée can be frozen up to 6 months.

5. To reheat, place frozen ramekins or custard cups in a 375°F oven for about 30 minutes, or until heated through.

6. To serve, dot with the 1 to 2 tablespoons butter and sprinkle with chopped parsley.

Yield: Six 6-ounce ramekins or custard cups

Zucchini Purée

*5 pounds zucchini, cut
    in 1/2-inch slices*
*5 tablespoons butter or
    margarine*
*1 1/2 teaspoons dried
    thyme*
    Salt and pepper

1. In a large saucepan, cook zucchini slices in enough water to cover for 10 minutes, or until tender when pierced. Drain.

2. In a food processor or blender, whirl zucchini, butter, and thyme until smooth. Season with salt and pepper to taste.

3. Freeze in plastic containers. Or spoon into ice cube trays, tray-freeze until firm, and turn out into plastic freezer bags. Purée can be frozen up to 6 months.

4. To serve, thaw purée in refrigerator; reheat in saucepan or microwave oven.

Yield: 36 cubes or four 8-ounce containers

Guide to Freezing Vegetables

This guide gives times for water-blanching and, if appropriate, steam-blanching. (At altitudes above 5,000 feet, add 1 minute to water-blanching and 2 minutes to steam-blanching time.) If a time range is given, blanch smaller vegetables for the shorter time; larger vegetables for the longer period. For water-blanching, allow 1 gallon water for each pound of vegetables. Begin to count blanching time when water returns to boil. If this takes more than 1 minute, reduce amount of vegetables in next batch. Unless otherwise noted, cool vegetables in ice water or under cold running water. See page 22, "How to Protect Fruit Color," for information on ascorbic acid. Vegetables can be stored in the freezer 8 to 12 months. Avoid refreezing thawed vegetables; there will be a noticeable loss of quality.

Artichokes
For whole artichokes, remove coarse outer leaves; cut 1 inch off top; trim tips and stem. Water-blanch 10 minutes, adding 1/2 cup lemon juice or 1 tablespoon ascorbic acid to each 2 quarts water. Cool. Pack in plastic freezer bags.

Asparagus
Break off woody ends; peel if desired. Sort according to size. Water-blanch 2 to 4 minutes. Steam-blanch small stalks 3 minutes; large stalks 4 1/2 minutes. Cool and pack.

Beans
Lima: Hull beans. Water-blanch 2 to 4 minutes. Steam-blanch 3 to 6 minutes. Cool and pack.
String or green: Remove ends. Cut into 1- to 2-inch pieces or slice lengthwise French style. Water-blanch 2 to 3 minutes. Steam-blanch 3 to 4 minutes. Cool and pack.

Beets
Scrub and peel; leave whole if small or slice or dice if large. Water-blanch whole beets 5 minutes; sliced or diced beets 3 minutes. Steam-blanch 7 and 4 minutes, respectively. Cool and pack.

Broccoli
Trim outer leaves. Slice stalks crosswise into thin rounds. Separate top into 1 1/2-inch or smaller flowerets. Water-blanch 3 to 4 minutes. Steam-blanch 5 minutes. Cool and pack.

Brussels sprouts
Remove coarse outer leaves. Sort according to size. Water-blanch 3 to 5 minutes. Steam-blanch 4 to 7 minutes. Cool and pack.

Cabbage
Remove outer leaves. Cut heads into wedges of convenient size. Water-blanch 3 to 4 minutes. Steam-blanch 4 to 6 minutes. Cool and pack.

Carrots
Remove tops and scrape. Freeze small, tender carrots whole; cut others into cubes, slices, or lengthwise sticks. Water-blanch whole carrots 5 minutes; pieces 2 minutes. Steam-blanch 7 and 3 minutes, respectively. Cool and pack.

Cauliflower

Break into 1-inch flowerets. Water-blanch 3 minutes, adding 1 tablespoon salt or vinegar to each gallon of water to retain color. Cool and pack.

Celery

Remove leafy tops and coarse strings. Cut into 1-inch pieces. Water-blanch 3 minutes. Steam-blanch 4 minutes. Cool and pack.

Celery root

Cut away leaves and root fibers. Scrub thoroughly. Cook until almost tender, about 20 to 30 minutes. Cool. Peel and slice or dice. Pack.

Corn

Husk; remove silk. Sort according to size. Pierce each cob lengthwise through its center with ice pick or sharp knife. Water-blanch 3 to 5 minutes. Steam-blanch 4 to 7 minutes. Cool. Pack ears, or cut kernels from cob and pack.

Eggplant

Peel and slice or dice. Water-blanch 4 minutes. To retain color, dip in a solution of ½ cup lemon juice or 1 tablespoon ascorbic acid to 5 cups water after blanching. Cool and pack.

Ginger root

Do not blanch. Wrap whole, uncut root in freezer packaging material. To use, grate or slice *frozen* root. Immediately return unused portion to freezer.

Grape leaves

Water-blanch 1½ minutes. Cool and pack.

Kohlrabi

Select young, tender roots. Cut off tops; peel and dice. Water-blanch 1 to 2 minutes. Steam-blanch 2 to 3 minutes. Cool and pack.

Mixed vegetables

Prepare and blanch each vegetable separately, according to directions for that vegetable. Cool and package together.

Mushrooms

Cut off ends of stems. Sort according to size. Leave whole or slice. Water-blanch 2 to 4 minutes, adding 1 tablespoon lemon juice to each quart water. Cool and pack. Or sauté in butter or margarine until tender. Cool to room temperature and pack.

Okra

Select young, tender green pods that snap easily. Remove stems, but do not cut open seed cells. Water-blanch 3 to 4 minutes. Steam-blanch 4 to 6 minutes. Cool. Leave whole or slice crosswise and pack.

Onions

Peel; chop or slice. Do not blanch. Spread in single layer on baking sheets; tray-freeze (see page 14) until firm. Pack. Use within 2 months.

Parsnips

Cut off tops. Peel and cut into ½-inch cubes or slices. Water-blanch 2 minutes. Steam-blanch 3 minutes. Cool and pack.

Peas

Green: Shell peas. Water-blanch 1½ minutes. Steam-blanch 2½ minutes. Cool and pack.

Chinese or edible-pod: Remove stem and blossom ends and any strings. Water-blanch 1½ to 2 minutes. Steam-blanch 2 to 3 minutes. Cool and pack.

Peppers

Chile (Anaheim, poblano): Roast peppers over an open flame or under the broiler until skin is charred and blistered. Place peppers in a plastic bag; set aside for 5 to 10 minutes. Peel peppers, cut in half, and remove seeds. (To keep the oils in peppers from irritating your skin, wear rubber gloves.) Package in layers, separating layers with waxed paper.

Sweet green or red: Remove stem and seeds; slice or dice. Do not blanch. Spread in single layer on baking sheets; tray-freeze (see page 14) until firm. Frozen peppers lose their crispness when thawed. (Use in cooked dishes.)

Potatoes, cooked

Baked and stuffed: Bake and stuff as usual. Cool. Tray-freeze (see page 14) on baking sheets until firm. Wrap individually in foil; then pack in plastic freezer bags. Loosen foil and reheat. Use within 1 month.

Mashed: Prepare as usual. Cook and pack. Use within 2 weeks.

Pumpkin

Cut into pieces, removing seeds and fibrous material. Steam or bake until tender. Scoop pulp from skin and mash. Cool and pack.

Spinach and other greens

Wash thoroughly to remove sand and grit. Water-blanch leaves 1½ minutes, stirring well. Cool and pack. Greens may be chopped before packing if desired.

Squash

Summer varieties: Cut into ½-inch slices. Water-blanch 3 minutes. Steam-blanch 4 minutes. Cool and pack.

Winter varieties: Halve; remove seeds and fibrous material. Cut into chunks. Steam or bake until tender. Scoop pulp from skin and mash. Cool and pack.

Sweet potatoes

Cook until almost tender. Cool at room temperature. Leave whole, halve, slice, or mash. Dip halves or slices in a solution of ½ cup lemon juice or 1 tablespoon ascorbic acid to 1 quart water for 5 seconds. To keep mashed sweet potatoes from darkening, add 2 tablespoons orange or lemon juice to each quart of mashed potatoes. Pack.

Tomatoes

Freeze only as sauce, paste, or purée. Prepare as usual; cool. Pack in containers.

Top: Chop chives and package in freezer containers, jars, or bags. Bottom: Wash herbs and pat them completely dry with paper towels before packaging in small bags.

How to Prepare and Pack Herbs

Herbs add a heady aroma and a pungent flavor to an infinite array of dishes. You can grow your own indoors or out, and major supermarkets now feature fresh herb displays in the produce department.

Because the oils in delicate herbs like parsley, chervil, basil, and chives are very volatile, freezing preserves their flavor better than drying. Full-flavored herbs like rosemary, sage, and oregano can be frozen or dried. See page 84 for information on drying herbs.

It's easy to freeze herbs for year-round enjoyment. The ideal time to harvest them is in the morning, before the sun's heat has warmed them. Most herbs are best picked at the peak of growth, just as the plant is about to flower. At this point, the fragrant oils that give flavor to the leaves are abundant. Parsley, chervil, and savory are the exceptions—they should be harvested when the leaves are young and tender.

It's easiest to cut entire stems or branches of most herbs. If you want the plant to continue its growth, leave some foliage at the base.

After harvesting, wash herbs in cold water; remove any deteriorating leaves. Drain and pat completely dry on paper towels. Strip leafy herbs from their stems. Package, whole or chopped, in convenient-size containers or small plastic bags. Group small bags together in a large plastic bag, labeling each bag with the name of the herb it contains. Be sure to remove as much air as possible during packaging.

Frozen herbs darken and wilt when thawed, so add them to soups, sauces, and salad dressings while they're still frozen. You can substitute frozen herbs for fresh in any recipe. They keep up to a year in the freezer.

Herb butters are a great way to use fresh herbs while they're at their peak; the butters also freeze beautifully. They add a flavorful twist to vegetables, breads, and rolls, or use them to baste grilled fish, shellfish, and poultry. Recipes for Herb Butters are at right.

Pesto Sauce

Make pesto in the summer and freeze it to toss with hot or cold pasta or to flavor minestrone soup.

- 4 to 5 cups firmly packed fresh basil leaves
- 1 cup parsley sprigs
- 5 to 6 cloves garlic
- 2/3 cup pine nuts or walnuts, lightly toasted
- 1 cup freshly grated Parmesan cheese
- 1 to 1 1/2 cups olive oil
  Salt
  Ground black pepper

Place basil, parsley, garlic, nuts, Parmesan cheese, and 1 cup of the oil in a food processor or, in batches, in a blender. Blend to a smooth paste, adding more olive oil if necessary. Season to taste with salt and pepper.

Yield: About 2 1/2 cups

To freeze: Spoon into small freezer containers and freeze up to 1 year.

Herb Butters

Tightly wrapped, frozen herb butters keep for 6 months. Simply slice off the amount you need. Other flavoring possibilities include orange, strawberry, prune, and almond.

Lemon-Parsley Butter

- 1/2 cup (1/4 lb) butter, softened
- 2 tablespoons finely chopped parsley
- 1/2 teaspoon minced garlic
- 2 to 3 teaspoons lemon juice
- 1 teaspoon finely grated lemon peel
  Dash salt and ground white pepper

Beat butter with an electric mixer or by hand until

soft and light. Beat in parsley, garlic, lemon juice, lemon peel, and salt and pepper. Form into a roll and wrap in freezer wrapping material. Freeze up to 6 months.

Dill Butter

1/2 cup (1/4 lb) butter, softened

1 to 2 tablespoons finely chopped fresh dill or 1 teaspoon dried dillweed

2 teaspoons lemon juice

1/2 teaspoon finely grated lemon peel

Dash salt and ground white pepper

Beat butter with an electric mixer or by hand until soft and light. Beat in dill, lemon juice, lemon peel, and salt and pepper. Form into a roll and wrap in freezer wrapping material. Freeze up to 6 months.

Lime-Cilantro Butter

1/2 cup (1/4 lb) butter, softened

1 to 2 tablespoons finely chopped cilantro

3 to 4 teaspoons lime juice

1/2 teaspoon finely grated lime peel

Dash salt and ground white pepper

Beat butter with an electric mixer or by hand until soft and light. Beat in cilantro, lime juice, lime peel, and salt and pepper. Form into a roll and wrap in freezer wrapping material. Freeze up to 6 months.

Fresh basil, thyme, dill, sage, chives, and rosemary can be stored in the freezer up to one year.

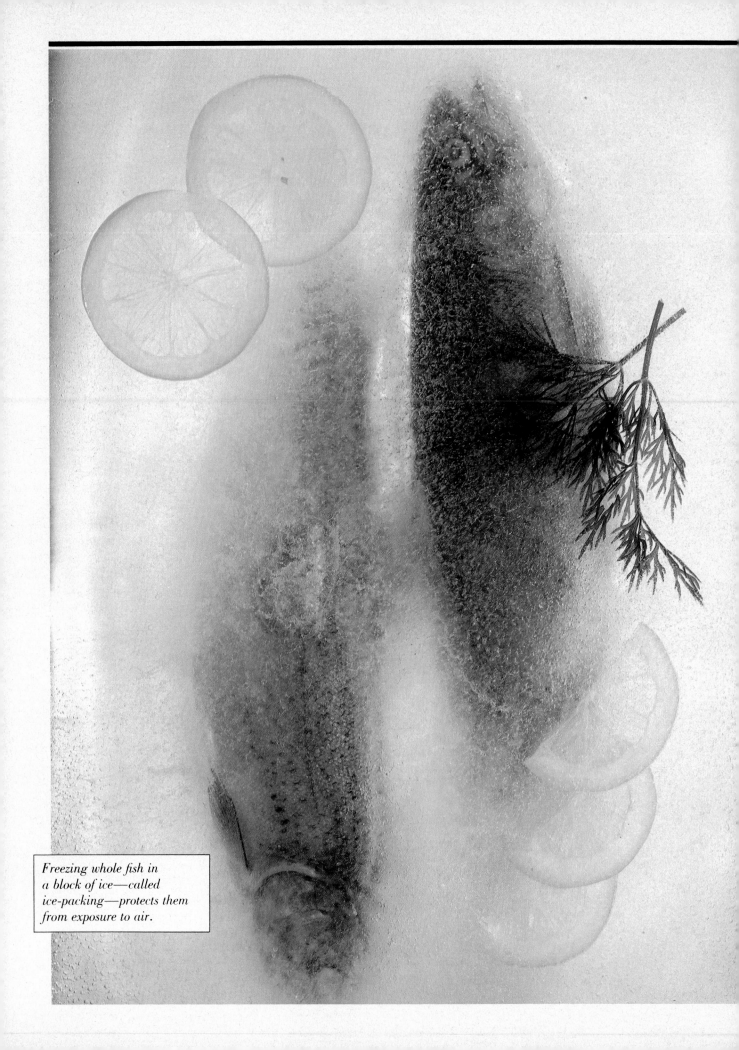

Freezing whole fish in a block of ice—called ice-packing—protects them from exposure to air.

Freezing Meat, Poultry, Fish & Main Dishes

I n terms of dollars and sense, meat, poultry, and fish may be your most valuable frozen assets. Freezing is the best way to retain the good taste and texture of these foods. Before you calculate how much and what kinds of meat, poultry, or fish to buy, decide how much freezer space you want to devote to these foods and consult the storage time-tables throughout this chapter to determine how fast you must turn them over to preserve quality.

There are several ways to economize when purchasing meat, poultry, or fish for the freezer. Shopping for specials can add up to significant savings, and remember that meat, poultry, and fish have seasonal peaks. Beef, for example, is more abundant in the fall, turkey is plentiful and less expensive during the holidays, and many species of fish are available only a few months of the year. You can also save by purchasing primal cuts of meat (such as loin, rib, round, or chuck) or whole birds and cutting them up yourself into meal- or serving-size portions.

With a freezer, shopping for meat, poultry, and fish becomes an occasional pleasure rather than a daily chore. And leftovers can be saved for use in soup stocks, casseroles, or sandwiches.

Purchase pork chops when they're on special; then package a meal's worth and freeze.

MEAT

Americans consume an average of 77 pounds of beef each year. And meat in general—beef, pork, lamb, veal—accounts for a significant amount of the food dollar spent. It's no wonder that many people buy and freeze meat when it's on special or that some families purchase a side or a quarter carcass.

Before you rush out and buy a carcass, evaluate your needs, available freezer space, preferred meat cuts, and budget. Although the price may sound enticing, the cost per pound of bulk meat is based on the gross, untrimmed weight. Trimming removes at least 25 percent of the gross weight. You may have several choices as to how the side is butchered, but be aware that there will be many more packages of stew and ground meat than of roasts and steaks. A side of meat will also consume much of your freezer space. (A side of beef, for example, weighs 250 to 400 pounds.) If you want to freeze other foods, too, perhaps shopping for meat throughout the year makes more sense. Finally, consider the quantity of meat your family can use within the safe storage period. (See the chart opposite for recommended storage times.) If you decide to invest in a carcass, arrange for the supplier to quick-freeze the meat for you. Home freezers cannot freeze large quantities of food quickly at one time.

Venison and other large game can be handled in the same way as a side of beef. If you hunt great distances from home, find out about having the animal butchered, packaged, and frozen near the hunting area and bring it home on dry ice. Or, if you hunt closer to home, have the game cut, wrapped, and frozen by a commercial locker plant in your area. Consult the department of fish and wildlife for hunting restrictions in your state. Your county extension home economist may have additional game-preserving information.

PACKAGING AND FREEZING MEAT

Meat requires very little preparation for freezing. The main consideration is to freeze it as soon as possible after purchase to ensure freshness and top quality during storage. Because frozen meats are most useful when they are "ready to cook," trim excess fat and remove bones if practical, or cover sharp, extruding bones with folded freezer paper or foil so that they will not pierce the outer wrapper. Divide large cuts into meal-size pieces and wrap individually. Shape ground beef into patties. Cube cuts that you plan to use for stew. Do not season meat before freezing because this shortens freezer life.

Package meat in heavy-duty aluminum foil or freezer paper using either the butcher or the drugstore wrap method. (See page 13.) Separate a meal's worth of steaks, chops, and other individual cuts with double sheets of wrapping material before stacking and wrapping. Tray-freeze (see page 14) beef cubes, meatballs, chops, or ground meat patties and store in plastic bags. They can be conveniently removed in the quantity you need.

Freeze meatballs until firm; then package them together in plastic bags and use as needed. Separate hamburger patties with two pieces of waxed paper so that they can be thawed and cooked individually.

You can freeze prepackaged meat as it comes from the self-service meat counter in the supermarket for up to two weeks without rewrapping. For longer freezer storage, repackage the meat in a suitable freezer wrapping material. Butchers in meat markets sometimes wrap meat in coated freezer paper that can go directly from store to freezer. Uncoated butcher paper, however, is not a suitable wrap for freezing.

Just about any cooked meat can be frozen—the leftover beef roast from dinner, half of that succulent roasted leg of lamb, or country-style pork spareribs from the barbecue. Used within two to three months, leftover cooked meat can be transformed into economical and convenient meals—a tasty hash, spicy lamb curry, or crispy pork fried rice.

Leftover crumbled cooked bacon and finely chopped cooked ham can be frozen wrapped in aluminum foil for several weeks to top salads, casseroles, or vegetables. Carefully wrap ham bones in freezer material for split pea soup.

Be sure to label each package with the date, type of cut, and weight or number of servings. If you wrap in freezer paper or aluminum foil, you won't be able to see the contents.

Thawing Meat

The safest and best way to thaw meat to retain optimum quality and texture is in the refrigerator. Although thawing takes substantially longer than at room temperature, the meat will retain its original flavor and moistness, and there will be less chance of spoilage. Thaw in the original freezer wrapping on a dish to catch any drippings. Allow 4 to 6 hours per pound for thawing.

The microwave oven is the ideal chamber for thawing meat quickly, safely, and with quality results. A 4-pound standing rib roast will defrost in 60 minutes in a microwave oven while taking 24 hours or more in the refrigerator. Consult your oven manual for exact defrosting times and techniques.

Cooking Frozen Meat

Most cuts of meat can be cooked without thawing; however, large cuts like roasts may dry out, lose more juices, and overcook on the outside before the center is done. A frozen roast will take $1\frac{1}{3}$ to $1\frac{1}{2}$ times the usual cooking period. Use a meat thermometer for accuracy. Frozen ground meat, chops, and steaks take $1\frac{1}{2}$ to 2 times longer to cook. Frozen cuts cooked by moist heat need little additional cooking time, if any.

Meats that are to be shaped, stuffed, coated with flour or crumbs, dipped in batter, or deep-fried should be completely thawed first.

If the market where you shop buys "boxed beef," you may be able to get a good buy on sub-primals—small wholesale units that are usually boneless—particularly if you cut them up yourself.

Timetable for Storing Meat

Type of meat	Storage time at 0°F*
Beef	
Roasts, steaks	6 to 12 months
Beef for stew	3 to 4 months
Ground beef	3 to 4 months
Beef variety meats (liver)	3 to 4 months
Corned beef, bologna, luncheon meats†	2 weeks
Lamb	
Roasts, chops, cubes	6 to 9 months
Ground lamb	3 to 4 months
Pork	
Roasts, chops, ribs	3 to 6 months
Ground pork	1 to 3 months
Ham†,‡	2 months
Bacon, frankfurters†	1 month
Veal	
Roasts, chops, cutlets, cubes	6 to 9 months
Ground veal	3 to 4 months
Sausage†	1 month
Cooked meat	2 to 3 months

*Foods stored longer than the recommended period will suffer a loss of quality but will still be safe to eat. Storing foods at temperatures higher than 0°F shortens the storage period considerably.

†Processed, cured, smoked, and ready-to-serve meat products do not retain their quality long in the freezer because their high fat content and seasonings accelerate rancidity and promote changes in flavor and texture. If it is necessary to freeze these meats, use them within the times indicated.

‡Do not freeze *canned* hams and *canned* picnics.

Plastic freezer bags are well suited for packaging whole birds, fish, and all cuts of meat. Be sure to cover any sharp, protruding bones with folded freezer paper or foil so that they will not pierce the freezer bag.

Poultry

Pound for pound, poultry is one of the most economical sources of animal protein. Stock up when poultry is a good buy; then freeze. Whole birds are usually the most economical. The larger the chicken or turkey, the greater the ratio of meat to bone and the better the value. But don't overlook birds like Cornish hens, capon, quail, pheasant, and duck. Once considered restaurant-only fare, today they are available frozen in many supermarkets and fresh at some specialty markets and butcher shops.

Fresh poultry sold in self-service meat departments must be repackaged before freezing. Simply remove the original wrap, and rinse and dry the bird. Package following the instructions below. (Be sure to remove the giblets before repackaging whole birds.) Commercially frozen poultry, such as turkey, capon, and Cornish hens, is already packaged for freezer storage and needs no additional wrapping.

When you freeze fresh poultry at home, the flesh sometimes turns dark around the bone. This is a result of blood and bone marrow leaching out of the bone due to slow freezing. It is normal and harmless.

Package and freeze small game, such as rabbit, according to the directions for poultry.

Packaging and Freezing Poultry

Whole birds. Freeze poultry unstuffed; home-prepared stuffing can develop harmful bacteria if frozen inside the bird. Package the giblets and liver separately; they develop an off-flavor when packaged with the bird. Tie the ends of the legs together and press the wings tightly to the body. Wrap in freezer paper or heavy-duty aluminum foil using the butcher or the drugstore wrap method illustrated on page 13. Or package in a large, heavy-duty freezer bag and water-dip as described on page 14.

Pieces. Package poultry pieces in convenient combinations: cut-up fryers for southern fried chicken; chicken breasts for chicken piccata; legs and thighs for barbecuing; wings for appetizers; backs and necks for chicken stock. Separate and tray-freeze pieces (see page 14) on a baking sheet until firm. Then package in plastic freezer bags so that you can remove individual pieces easily. If you are wrapping several pieces in freezer paper or aluminum foil, place close together and package tightly to eliminate as much air as possible. Or put poultry pieces into a freezer bag and expel the air using the water-dip or the straw method. (See pages 12 and 14.)

Cooked poultry. Leftover cooked poultry can be frozen successfully. To save space, remove the meat from the bones and package in meal- or recipe-size quantities.

When packaging either raw or cooked poultry, be sure to label each package with the exact contents, date frozen, and weight or number of pieces or servings.

Thawing Poultry

For even cooking, thaw whole poultry before you cook it. Poultry pieces may be cooked without thawing, but you'll need to increase the cooking time by a third to a half, and the poultry may be somewhat drier than usual.

It's safest to thaw poultry in the refrigerator or to defrost it in a microwave oven. If you are in a hurry, however, it's safe to thaw poultry in a basin of cold water.

The best way to thaw turkey is in its original wrapping in the refrigerator, which takes about 24 hours for each 5 pounds of weight. If you forget to take the

Timetable for Storing Poultry

Type of poultry or game	Storage time at 0°F*
Chicken, capon, turkey, Cornish hens, small game	
Whole	12 months
Pieces	6 to 9 months
Giblets	2 to 3 months
Goose, duck, game birds	
Whole	6 months
Cooked poultry	
Pieces not in broth or gravy	1 month
Pieces in broth or gravy	6 months
Cooked poultry dishes (casseroles, stews)	3 to 6 months

*Foods stored longer than the recommended period will suffer a loss of quality but will still be safe to eat. Storing foods at temperatures higher than 0°F shortens the storage period considerably.

turkey out of the freezer in time, thaw it in cold water, changing the water every 30 minutes. This method takes about half an hour per pound of turkey. Refrigerate or cook the turkey as soon as it is thawed.

FISH AND SHELLFISH

There's nothing more satisfying for a weekend fisherman than to bring home a bounty of fresh fish. Salmon grilled on the barbecue, trout sautéed in lemon butter, or striped bass baked with herbs are especially delicious when the fish are freshly caught. But what do you do with the six remaining trout? Freeze them!

Properly handled from the moment it is pulled from an ocean, lake, or stream until it is placed in the freezer, frozen fish is nearly as good as its fresh counterpart. Handling and timing are key, because fish and shellfish are the most perishable of all fresh foods. Most fishermen recommend that you freeze fish within 24 hours of catching it. In between, *keep it on ice or under refrigeration*.

If you don't have a fisherman in your family or if you want to buy and freeze fresh salmon, for example, when it's in season, buy the freshest fish you can find. When shopping for fish, observe its storage conditions. Does the retailer keep the work area clean and uncluttered? Is the fish held on ice or under refrigeration? Ask your retailer when and where the fish was caught.

A fresh fish has firm flesh; bright, clear eyes; scales that are shiny and tight to the skin; and gills that are reddish and free of slime. Fresh fish has a mild, sweet odor; a strong "fishy" odor means that the fish is old.

Clean and prepare fish for the freezer exactly as you would if you were going to cook it immediately, or have the retailer clean the fish for you. How you handle it after that will depend on whether it is a lean or a fatty fish. Lean fish can be stored frozen for up to 6 months, while fatty fish are best consumed within 3 months. This is because the fat in fish becomes rancid in a relatively short time. Fatty fish include mackarel, most types of salmon, bluefish, smelts, herring, trout, shad, and tuna. Lean fish include sole, cod, bass, haddock, halibut, red snapper, and yellow pike.

Fatty fish will taste fresh longer and will be less likely to turn rancid if treated

briefly in an ascorbic acid mixture before freezing. Dip each fish in a solution of 2 teaspoons ascorbic acid to 1 quart cold water for 30 seconds; drain. (See page 22 for information on purchasing ascorbic acid.) Lean fish keep best if dipped for 30 seconds in a brine of ⅔ cup salt to 1 gallon cold water.

PACKAGING AND FREEZING FISH AND SHELLFISH

Whole fish. To freeze whole fish, first *ice-glaze* for added protection against oxidation and moisture loss. After the fish has been drawn (scales, gills, and entrails removed), dip it in a brine or ascorbic acid solution as explained earlier;

Heavy-duty foil molds tightly around awkwardly shaped whole birds.

"Previously frozen" or "fresh frozen" fish is delivered to supermarkets frozen, unwrapped at the store, and set in the seafood case to thaw gradually. Refreezing is possible, but the quality will be much better if you eat the fish right away. If you refreeze previously frozen seafood, do it without delay. Seafood that has been thawed for more than 12 hours or that has been held in the refrigerator or on ice is a haven for bacterial growth. Refreezing may be risky.

then tray-freeze the fish (see page 14) on a baking sheet until it's completely frozen. Next, dip briefly in a bath of ice water. A glaze will form immediately. Return the fish to the freezer to solidify the glaze. Repeat the dipping and freezing as many times as necessary to build up a layer of ice at least ⅛ inch thick. Wrap in heavy-duty foil, freezer paper, or a large plastic freezer bag and freeze.

Another packaging method unique to fish is *ice-packing*. Although it requires more freezer space, ice-packing is a desirable method because it protects the fish from air. Place the fish in a washed milk carton lined with a plastic freezer bag or in a plastic container large enough to hold the fish. Fill the bag or container with cold water to within ½ inch of the top, making sure the fish is totally submerged. Seal tightly and freeze.

Small whole fish, such as smelts or fresh sardines, may be ice-packed or simply tray-frozen (see page 14) and then wrapped in freezer paper, heavy-duty aluminum foil, or a freezer bag.

Fillets and steaks. Before packaging, dip fillets and steaks in either a brine or an ascorbic acid solution, as explained earlier. Wrap pieces individually or stack them, separating the pieces with two sheets of freezer wrapping material. Wrap tightly, following either the butcher or the drugstore wrap method illustrated on page 13. Then overwrap with freezer wrapping material to protect the freezer against any fish odor. You also may ice-pack the fish pieces, as explained under "Whole Fish."

Shellfish. Because of its delicate nature, shellfish is best consumed fresh. However, if you have an abundance, freezing may be your best option.

Crab and lobster are best frozen after they are cooked. Freeze them in the shell, or remove the meat and freeze it separately. Shrimp, mussels, clams, oysters, and scallops are best frozen uncooked. Remove the heads from shrimp, tray-freeze (see page 14), and package in plastic freezer bags. Scallops can also be tray-frozen. Shuck clams, oysters, and mussels, saving their juice; strain the juice to remove any sand. Rinse the shellfish and pack in rigid containers. Cover the shellfish completely with their juice and additional brine made of 1 tablespoon salt to 1 quart water. Leave ½-inch headspace and seal tightly.

PURCHASING AND HANDLING FROZEN FISH

When purchasing frozen fish, select solidly frozen packages. Avoid broken packages, dehydrated packages with a honeycomb appearance and an off-color, and packages with a pronounced odor. When you take home seafood from the frozen food case, leave it in its original packaging and freeze immediately.

THAWING FROZEN FISH

Because thawing causes the delicate flesh of fish and shellfish to break down, you'll get the best results by cooking most frozen fish and shellfish before they thaw completely. For ice-glazed or ice-packed fish, thaw just enough to remove the outer coating of ice. Thaw fillets and steaks until they can be separated or cut into serving-size pieces. Allow additional cooking time as necessary. Fish and shellfish firm up and become opaque when done.

There are a few exceptions to the rule. Prepackaged breaded or battered frozen fish items should not be thawed at all before cooking. Fish that is to be breaded, battered, deep-fried, or stuffed is easier to handle when the fish is thawed first.

Completely or partially thaw fish in the refrigerator. When you must thaw the product more quickly, leave it in its original wrapper and place it in a plastic bag; then submerge the bag in a basin of cold water. Check periodically, and cook or refrigerate the fish as soon as it is thawed. Or thaw fish quickly in a microwave oven.

TIMETABLE FOR STORING FISH AND SHELLFISH

Type of fish or shellfish	Storage time at 0°F*
Fatty fish (tuna, salmon, etc.)	1 to 3 months
Lean fish (haddock, sole, etc.)	4 to 6 months
Clams and scallops	3 to 4 months
Crab and lobster	1 to 2 months
Oysters	1 to 3 months
Shrimp	4 to 6 months

*Foods stored longer than the recommended period will suffer a loss of quality but will still be safe to eat. Storing foods at temperatures higher than 0°F shortens the storage period considerably.

For optimum quality, freeze your catch within 24 hours of pulling it from the water.

Stockpile beef, chicken, and fish bones in the freezer to make homemade stock.

MAIN DISHES TO MAKE AHEAD AND FREEZE

Do you live in a household where busy schedules call for meals at varying hours? Have little time in the evenings to prepare a balanced dinner? Live alone or have a small family? Entertain large groups frequently? Frozen main dishes—stews, casseroles, soups, pasta sauces—can answer these meal-planning challenges.

When you prepare a favorite dish, make a double batch and package the second in serving sizes for future meals. Commercially frozen TV dinners can be an inspiration to planning ahead. The next time you prepare a meal, plan extra servings of each course; then package individual meals in pie plates or divided foil or plastic plates.

Almost any casserole or main dish can be frozen. The choice of container depends on the food and the method of reheating. A good trick for casseroles is to line the baking dish with heavy-duty aluminum foil, fill it with food, and freeze until firm. Then remove the food from the dish. This frees your baking dishes for other uses and makes for easy cleanup, too. See page 13 for casserole-wrap instructions.

Package foods for microwave defrosting and heating in ceramic, glass, or plastic containers. Match the size of the container to the amount of food you want to freeze. Filling the container (while allowing adequate headspace) keeps air out and extends storage time. Main dishes keep from 3 to 6 months in the freezer.

Rely on the freezer to make entertaining easy. Select recipes that can be prepared in advance and frozen, such as Caribbean Black Bean Soup and Twister Chili (pages 44 and 45). Both go directly from freezer to stove or microwave. Reheating will be easier and faster if you freeze the chili or the soup in meal- or serving-size portions.

Avoid freezing fried foods because their crisp coatings become soggy. Crunchy casserole toppings, such as cracker or bread crumbs, will also turn soggy in the freezer, so add them during reheating. Other ingredients to avoid in frozen main dishes are uncooked potatoes, which become mushy and may darken, and hard-cooked egg whites, which become tough.

Many frozen main dishes can be reheated in the oven or in a double boiler after they have thawed enough to permit removal from their freezer container. A microwave oven is ideal for both defrosting and reheating main dishes. Consult your oven's manual for defrosting and heating times.

Beef Stock

There is nothing like homemade beef stock for soups and sauces. Spend the time to make a double batch, and don't be caught short.

- 1/4 cup vegetable oil
- 2 pounds beef bones, sawed in 3-inch pieces A 5-pound beef shin bone, cut in 2- to 3-inch pieces
- 2 medium onions, unpeeled and quartered
- 2 carrots, cut in 2-inch pieces
- 1 celery stalk, cut in 2-inch pieces
- 3 tablespoons all-purpose flour
- 5 quarts water
- 2 cloves garlic
- 4 sprigs parsley
- 2 sprigs fresh thyme or 1/2 teaspoon dried thyme
- 1 bay leaf
- 1/2 teaspoon whole black peppercorns

1. Heat oil in a large roasting pan. Add bones and roast in 400°F oven until browned (15 to 20 minutes). Add onions, carrots, and celery to pan and roast until they begin to brown. Sprinkle bones and vegetables with flour and continue roasting until lightly browned.

2. Remove bones and vegetables from roasting pan and place in large stock pot. While roasting pan is still hot, add 1 cup of the water to pan and scrape up browned bits on bottom. Pour into stock pot.

3. Cover bones with at least 2 inches of water. Add garlic, parsley, thyme, and bay leaf. Bring to a boil over medium-high heat. Boil 10 minutes, skimming scum as it rises to the top. Reduce heat to barely simmering. Do not let stock boil again. Cook, uncovered, about 7 hours, skimming top occasionally. Add water as needed to keep bones covered. During the last hour of cooking, add peppercorns.

4. Strain stock through a colander lined with several layers of cheesecloth. Discard bones and vegetables. Cool stock to room temperature. Refrigerate overnight. Remove layer of fat that solidifies on top.

5. Return stock to heat; gently simmer 3 hours to reduce stock and concentrate flavor.

Yield: 2 1/2 to 3 quarts

To FREEZE: Cool stock to room temperature. Pour into freezer containers, leaving 1/2-inch headspace for pints; 1-inch headspace for quarts. Freeze up to 8 months.

Fish Stock

Fish stock, often called fish fumet, is wonderful to have handy for seafood sauces and soups. Save trimmings from fish and freeze, or ask for bones and trimmings at your fish market. Season stock with salt when you use it.

- 3 to 4 pounds fish trimmings, bones, heads, and tails
- 2 quarts water
- 2 cups dry white wine
- 1 large onion, chopped
- 1 stalk celery, chopped
- 1 lemon slice
- 5 sprigs parsley
- 2 sprigs fresh thyme or 1/2 teaspoon dried thyme
- 1 bay leaf
- 4 whole white or black peppercorns

1. Rinse fish trimmings and bones under cold running water. Remove gills from heads; crack large bones. Place trimmings and bones in a 6- to 8-quart nonaluminum pot. Add water and wine. Bring liquid to a boil and cook 5 minutes, skimming scum that rises to the top.

2. Reduce heat to a simmer. Add remaining ingredients (except peppercorns). Simmer stock for 20 minutes. Add peppercorns and simmer 10 minutes longer, skimming surface occasionally.

3. Strain stock through a colander or sieve lined with several layers of cheesecloth. Discard bones and vegetables.

Yield: 8 to 9 cups

To FREEZE: Cool stock to room temperature. Pour into freezer containers, leaving 1/2-inch headspace for pints; 1-inch headspace for quarts. Freeze up to 8 months.

Freezing Stocks and Sauces

The freezer is a wonderful storehouse for the makings of homemade stock. Collect beef, veal, chicken, turkey, or fish bones and store them in separate bags in the freezer until you've accumulated the amount you need. (See Beef and Fish Stock recipes at left.) Once you've made the stock, freeze it in pint or quart containers. What a convenience to have homemade stock handy to use as a base for soups, stews, and sauces. Once you've made your own stock, you'll never buy the commercially canned product again.

Some sauces separate when frozen. Reheating at a low temperature and stirring constantly will usually recombine a sauce. Pasta and rice tend to absorb liquid from their sauces during freezing and may turn soggy. To prevent this, undercook the pasta or rice slightly; it will finish cooking when the dish is heated. Or omit it from the recipe and add during reheating.

CARIBBEAN BLACK BEAN SOUP

Black beans, also known as turtle beans, make a wonderful, hearty soup that freezes beautifully.

- 1 pound dried black beans
- 3 tablespoons butter
- 3 medium onions, chopped
- 2 cloves garlic, minced
- 1 teaspoon dried oregano
- 1/2 teaspoon ground cumin
- 1/4 teaspoon dried thyme
- 1/8 teaspoon dried marjoram
- 1 bay leaf
  Pinch of ground coriander
  A 1 1/2- to 2-pound ham bone, trimmed of excess fat
- 1 quart beef stock
- 3 to 4 tablespoons dry sherry
- 4 canned tomatoes, drained and chopped or 1 cup peeled, chopped fresh tomatoes
  Salt and freshly ground pepper

  Garnishes:
- 3 hard-cooked eggs, chopped
  Fresh lime or lemon wedges
- 1/2 cup thinly sliced radishes
- 1/2 cup sour cream
- 6 green onions, chopped

1. Wash and sort beans. Place in a large pot, cover with cold water, and bring to a boil. Remove from heat and let stand, covered, 1 hour to rehydrate beans. Drain. Return beans to pot.
2. Melt butter in a large skillet over medium heat. Add onions and cook until soft and transparent. Add garlic and cook 1 minute.
3. To beans, add onion mixture, seasonings, ham bone, stock, and sherry. Partially cover and simmer 1½ hours.
4. Remove ham bone from pot and shred or dice meat. Discard bone.
5. Mash beans with potato masher, or coarsely purée soup in a blender or food processor. Add reserved ham and tomatoes to soup and cook 30 to 40 minutes longer. Season to taste with salt and pepper.*
6. Serve soup with an assortment of garnishes.

Yield: 6 to 8 servings

*TO FREEZE: Cool to room temperature. Pour into pint freezer containers, leaving ½-inch headspace. Freeze up to 6 months.

GREEK STEFADO

Here's a great one-pot dish. Serve half now and freeze the rest in meal-size freezer containers.

- 1/2 cup butter or margarine
- 3 large onions, thinly sliced
- 4 pounds lean beef, cut in 1-inch cubes
  Salt and freshly ground pepper
- 2 bay leaves
- 1/4 cup currants
- 6 to 10 ounces tomato paste
- 2 1/2 tablespoons red wine vinegar
- 3/4 cup dry red wine
- 1 1/2 tablespoons brown sugar
- 2 garlic cloves, minced
- 1/4 teaspoon ground cinnamon
- 1/4 teaspoon ground cumin
  Pinch ground cloves
- 1/2 pound Münster cheese, cubed
- 1 cup chopped walnuts

1. Melt butter in a large flameproof casserole. Sauté onions in butter until soft and translucent. Liberally season beef cubes with salt and pepper; add to onions.
2. Combine remaining ingredients (except cheese and nuts). Stir into beef mixture. Simmer, covered, about 2 hours, or until meat is tender. Adjust seasonings to taste.*
3. Scatter cheese and nuts over top. Bake, uncovered, in 350°F oven until cheese melts.

Yield: 8 to 12 servings

*TO FREEZE: Cool to room temperature. Spoon into pint or quart freezer containers, leaving ½- to 1-inch headspace. Freeze up to 4 months.

Add cheese and nuts when you reheat the dish.

CHICKEN FLORENTINE

Here's our version of "lean cuisine": a zesty chicken dish that's tasty and low in calories.

- 1 package (10 oz) frozen chopped spinach, thawed and squeezed dry
- 1 cup sliced mushrooms
- 1 clove garlic, minced
- 1/2 cup chicken broth
- 1/2 cup skim milk
- 1 1/2 tablespoons all-purpose flour
- 1 tablespoon dry sherry or vermouth
- 1/8 teaspoon white pepper
- 1/8 teaspoon Worcestershire sauce
  Dash of ground nutmeg
  Salt
- 1 cup diced cooked chicken
- 1 tablespoon chopped green onion
- 1/4 teaspoon finely grated lemon peel
- 2 teaspoons freshly grated Parmesan cheese

1. Spread spinach in bottom and up sides of a pie pan or small casserole. Set aside.
2. In a covered saucepan, simmer mushrooms and garlic in chicken broth for 5 minutes. Strain, retaining mushrooms, garlic, and broth. Set aside 10 of the nicest mushroom slices for garnish. Return ½ cup broth to pan.
3. Mix milk and flour together and add to pan. Stir over medium heat until mixture just begins to thicken.
4. Stir in remaining ingredients (except Parmesan cheese).

5. Spoon chicken mixture into pie pan. Sprinkle with cheese. Top with reserved mushrooms.*

6. Bake at 375°F until heated through (about 15 minutes).

Yield: 2 servings (about 300 calories each)

*TO FREEZE BEFORE BAKING: Cool to room temperature. Place plastic wrap directly on top of chicken. Wrap, and freeze up to 4 months. To heat, unwrap and place frozen casserole in 375°F oven for 30 to 35 minutes or until heated through.

TWISTER CHILI

This chili recipe feeds a crowd. Warm the tortillas, pour the cold beer, and you're ready for a twister party.

> 6 tablespoons bacon drippings or vegetable oil
> 6 pounds lean beef chuck, cut into 1/4-inch cubes
> 1 1/2 quarts beef broth
> 1 cup chili powder
> 1/4 cup honey
> 2 tablespoons ground cumin
> 2 tablespoons ground oregano
> 1 to 2 tablespoons chili pepper flakes
> 4 clove garlic, minced
> 4 cans (16 oz each) kidney beans, drained
> 2 cans (6 oz each) tomato paste
> 2 cans (16 oz each) stewed tomatoes
> Salt
> 1/4 cup cornmeal (optional)

1. Heat bacon drippings in a large flameproof casserole. Brown beef in batches, removing it as it browns. Return beef to casserole. Stir in broth, chili powder, honey, cumin, oregano, chili pepper flakes, and garlic. Partially cover and simmer 2 hours.

2. Add beans, tomato paste, and stewed tomatoes. Cook 40 minutes longer. Season to taste with salt. Add cornmeal, if desired, to thicken chili.

Yield: 16 to 20 servings

TO FREEZE: Cool to room temperature. Spoon into freezer containers, leaving 1/2-inch headspace for pints; 1-inch headspace for quarts. Freeze up to 6 months.

Make a large pot of robust Twister Chili. Serve one portion now and freeze the rest for another day's meal.

Pop frozen berry tarts or pies into the oven for a mouth-watering dessert without any fuss.

Freezing Baked Goods

With a freezer, you can stock up on store-bought breads and pastries to meet your needs for weeks or months and savor freshly baked flavor every day. And if you enjoy home baking, freezing lets you keep your family's favorite pies, cakes, cookies, and breads on hand for any occasion. When you consider that the most time-consuming parts of home baking are the preparation and cleanup, it makes sense to prepare double or triple the recipe, to make the most of your time. If you're about to bake an apple pie for Sunday's potluck barbecue, why not make three pies and freeze the other two unbaked for Sunday desserts over the next several months?

On the following pages, you'll find techniques for freezing specific types of baked goods. The chart on pages 50–52 is a guide to packaging, storing, and thawing baked goods by type. For information on freezer packaging materials and wrapping techniques to protect baked goods during storage, see pages 11–14.

Today's busy life-styles make the daily stop at the local bakery for bread, and perhaps a special cake or pastry, a thing of the past for many of us. Freezing can be the answer.

Separate waffles with double layers of waxed paper; then package together in plastic freezer bags.

Freezing Sandwiches, Canapés, and Pizza

A cache of sandwiches, fancy canapés, and pizza in the freezer is a time-saver for lunches, snacks, and parties. Frozen sandwiches packed in a lunch bag in the morning will thaw by noon. Remove canapés from the freezer about an hour before serving; arrange on trays to thaw.

Certain fillings and toppings freeze well: peanut butter (with or without jam); cheese spreads; cream cheese mixtures (with olives, dates, or raisins); and sliced meats, poultry, and canned tuna.

Mayonnaise and other oil-based dressings separate during freezing; substitute cream cheese, milk, catsup, or juice in fillings; or mix mustard or herbs with butter for a flavorful spread. (See Herb Butters, pages 32 and 33.) Add celery, carrots, radishes, hard-cooked egg whites, lettuce, and tomatoes to sandwiches or canapés after thawing.

Baked store-bought and homemade pizzas without processed meats like sausage and pepperoni freeze successfully. *Focaccia*, a treat that resembles a deep-dish pizza, also freezes well.

Breads and Rolls

Freezing can maintain the fresh taste and texture of bread and rolls for up to 6 months. You can freeze the baked product or, depending on the type of bread and the recipe, you can freeze the dough or batter before baking.

We have had good results freezing quick bread batter and then baking it without thawing first. In some cases you may have to increase the baking time slightly. See the recipes for Boston Brown Bread and Bran-Molasses Muffins on pages 53 and 54. You can freeze the muffin batter in paper-lined muffin pans, remove the liners from the pans, bag the muffins in plastic freezer bags, and then bake only as many as you need straight from the freezer.

All yeast breads and rolls freeze beautifully after baking, and many yeast doughs can be frozen before baking, too. If you wish to freeze before baking, most recipes will require additional yeast—between half a package and a package. Experiment with your favorite recipes or look for those designed to be frozen before baking. The recipe for Whole Wheat Bread (page 53) is one of these. Testing indicates that the dough freezes successfully for 3 weeks after it is shaped into loaves or rolls.

The makers of Fleischmann's active dry yeast also recommend freezing yeast dough immediately after it is kneaded. This eliminates the first rising, and no additional yeast is necessary. The dough may be stored in the freezer up to 6 weeks. When you wish to bake a loaf, thaw the dough and shape it. Then let the loaf rise until *tripled* in bulk before baking.

Fleischmann's has specially developed a number of yeast bread recipes that can be frozen before baking. To receive the recipes, write to Fleischmann's Yeast, Consumer Information Center, Nabisco Brands Plaza, Parsippany, NJ 07054.

Cakes and Cookies

Ideally, cakes should be frozen unfrosted and unfilled, and then frosted or filled after thawing. Powdered-sugar (buttercream-type) frostings are the only ones that freeze well. Egg white–based frostings separate or weep after freezing, as do whipped cream and custard fillings. If you must frost a cake before freezing, select a powdered-sugar frosting. A better approach is to prepare the cake and the buttercream frosting and freeze them separately. Package the frosting in a rigid plastic container and store it in the freezer for up to 2 months. Thaw in the refrigerator until spreadable.

Some experts do not recommend freezing cake batter; however, the Old-Fashioned Buttermilk Cake on page 54 was prepared and frozen in foil-lined cake pans. When firm, the batter was removed from the pans and the foil sealed tightly around it. After three weeks, the foil was peeled off and the frozen batter was returned to greased cake pans and baked immediately, without thawing. The cake was delicious. The Sour Cream Coffee Cake on page 54 is another example of a cake that can be batter-frozen and baked without thawing.

Just about any cookie, with the exception of meringue cookies, can be frozen, either baked or unbaked. Roll cookies, such as the Lockwood Butter Cookies on page 54, are simply sliced and baked when you're ready to serve them. Drop-cookie dough can be tray-frozen and then packaged in plastic food-storage bags. Take out and bake only as many cookies as you need. Most cookie recipes can be doubled or tripled; bake one batch and freeze the rest for the months ahead.

A precautionary note on flavorings and spices: The less expensive, imitation vanilla is not recommended for cakes and cookies that will be frozen. Further, some spices—such as cloves and nutmeg—intensify in flavor during freezing, so use slightly less than the amount called for in the recipe.

Pancakes and Waffles

If you have leftover pancake or waffle batter, cook it and freeze the finished product. Waffles and pancakes can be taken directly from the freezer and popped in a toaster to heat and crisp.

Pastry and Pies

To bake or not to bake pastry and pies before freezing is an often asked question. As with breads, the choice, in most cases, is yours. Many experienced pastry makers believe that freezing pastry and pies unbaked produces superior results to freezing them after baking. (A few pies *must* be baked before freezing, however; see "Freezing Baked Pies.") The option you choose will depend on available freezer space, type of pie, length of storage, and perhaps your time. (See page 58 for freezing quiche.)

Freezing Unbaked Pastry

Raw pastry can be frozen formed into balls to be rolled out after thawing; rolled flat; or shaped in metal, glass, ceramic, or aluminum foil pie or tart pans. Puff pastry can be frozen unbaked, before it is formed. Unbaked pastry has a shorter storage life than baked pastry.

Freezing Baked Pastry

You can form and bake pie and tart shells before freezing. These are useful for pies with fillings—chocolate, cream, lemon —that are not baked before serving. You can also freeze baked puff pastry shells.

Freezing Cream Puffs and Eclairs

Baked or unbaked shells for cream puffs and eclairs can be frozen. See the chart, page 51, for directions. Add the filling just before serving.

Freezing Unbaked Pies

Unbaked pies do not keep as long in the freezer as baked pies. If you plan to make several pies when blueberries, for example, are in season, freeze one pie unbaked to eat within several months; bake and freeze the other for winter enjoyment. See page 53 for a recipe for luscious Oregon Blueberry Pie.

When preparing fruit pies to freeze unbaked, increase the thickening agent in the filling, because frozen fruit gives off additional juice when baked. For each pie, increase cornstarch by ½ to 1 tablespoon; flour, by 1 to 2 tablespoons.

When freezing unbaked double-crust pies, do not cut steam vents in the top pastry until just before baking. This protects the filling from air. Pies with a lattice-work top crust have a shorter freezer life than those with a full top crust.

Freezing Baked Pies

Baked pies have a longer freezer storage life than unbaked pies.

Certain pies—mincemeat and nut pies with a sugar-egg base—should always be baked before freezing.

Freezing is not recommended for custard and cream pies; the filling becomes watery and tough.

Meringue toppings do not freeze well for more than several weeks. Over time, meringue shrinks and becomes tough. It's better to add freshly made meringue to the pie after thawing it.

Freezing Fruit Pie Fillings

To lengthen the freezer storage life of fruit pies (if you don't want to bake them before freezing), freeze the filling separately. See the chart, page 52, for directions.

Thawing Baked Goods

In general, frozen baked breads, fruit pies, cakes, cookies, and raw pastry dough can be thawed at room temperature. Cheesecakes; nut, pumpkin, and mincemeat pies; and puff pastry are best thawed in the refrigerator.

To thaw a frosted cake, remove the freezer wrapping first. Thaw all other baked goods in their wrappings. To prevent condensation, it's sometimes advisable to loosen the wrapping; see individual instructions in the chart on pages 50–52. If you plan to serve small baked items (such as rolls or muffins) warm, you need not thaw them; simply heat in a low oven, wrapped in aluminum foil. Waffles, pancakes, and bread slices may be toasted frozen. Thawing and warming times will vary depending on the density, size, and shape of the baked item.

Quick breads, yeast breads, cookies, baked pies, and cakes thaw quickly in a microwave oven on the defrost setting. If baked goods are thawed *completely* in a microwave oven, however, their edges may begin to cook and become dry. Let them complete their thawing during the "standing time." Consult the manufacturer's instructions.

Top: Cake and muffin batters can be frozen before baking. Line cake pans with foil for easy removal. Middle: Give pies additional protection by placing them in a bakery box after they've been wrapped for freezing. Bottom: Tray-freeze roll and drop cookie batters; then package together in freezer containers or bags.

GUIDE TO FREEZING BAKED GOODS

The following chart outlines preparation, packaging, freezing, and thawing instructions for homemade baked goods. The information for baked items also applies to store-bought baked goods.

BAKED GOOD	PREPARATION AND PACKAGING	FREEZER STORAGE TIME*	TO SERVE
BREADS AND ROLLS			
Biscuits	BAKED: Prepare and bake as usual. Cool. Wrap in freezer material. Or tray-freeze and store in plastic freezer bags; label.	2 to 3 months	Thaw in wrapping at room temperature or, if wrapped in aluminum foil, in a low oven.
Muffins	BAKED: Prepare and bake as usual. Remove from pans and cool. Wrap in heavy-duty foil. Or cool and freeze until firm in pans; remove from pans; store in plastic freezer bags; label.	6 to 12 months	Thaw according to instructions for Biscuits.
	UNBAKED: Prepare batter; spoon into paper-lined muffin pans; freeze until firm. Remove muffins from pans; store in plastic freezer bags; label.	3 weeks	Place frozen muffins in muffin pans and bake, adding about 5 minutes to baking time in recipe.
Nut and fruit breads, coffee cakes, and steamed breads	BAKED: Prepare and bake as usual. Cool. Wrap in freezer material and label.	2 to 4 months	Thaw according to instructions for Biscuits.
	UNBAKED: Prepare batter. Freeze until firm in baking pan. Wrap pan in freezer material; seal and label. Or line pan with aluminum foil (see casserole-wrap method, page 13); pour in batter; freeze until firm. Remove from pan, seal foil, label, and return to freezer.	3 weeks	Bake batter frozen, or thaw it. (If casserole-wrap method was used, peel off foil before returning batter to baking pan.) Additional baking time may be required if batter is frozen.
Yeast breads	BAKED: Prepare and bake as usual. Cool. Wrap in freezer material. Label.	6 to 8 months	Thaw according to instructions for Biscuits.
	PARTIALLY BAKED (HOMEMADE BROWN-AND-SERVE ROLLS): Prepare dough as usual. Bake at a temperature 100°F less than called for in recipe, until rolls are set but not browned. Cool. Wrap in freezer material. Label. NOTE: Not recommended for bread loaves.	3 months	Thaw in wrapping at room temperature. Remove wrapping and bake in hot oven until browned.
	UNBAKED: Follow recipe especially developed for freezing before baking. Or prepare dough as usual; knead; wrap dough in freezer material; label and freeze.	6 weeks	Remove wrapping and thaw at room temperature. Shape dough; let rise once until *triple* in bulk. (Omit first rising.) Bake according to recipe.
CAKES			
Angel, chiffon, and sponge	Prepare and bake as usual. Cool. If frosted, tray-freeze until firm; then wrap in freezer material. If unfrosted, wrap before freezing. For added protection, store in a box or rigid container, sealing edges with freezer tape. Label.	4 to 6 months	Loosen wrapping and thaw plain cakes at room temperature. Remove wrap from frosted cakes before thawing.
Butter and pound	BAKED: Prepare and bake as usual. Cool. Wrap and package according to instructions for Angel, Chiffon, and Sponge Cakes.	2 to 4 months	Thaw according to instructions for Angel, Chiffon, and Sponge Cakes.
	UNBAKED: Prepare batter. Line cake pans with aluminum foil (see casserole wrap, page 13); pour batter into pans and tray-freeze until firm. Remove cakes from pans; seal foil, label, and return to freezer.	2 months	Peel off foil and return batter to pans. Bake according to recipe, allowing additional time as necessary.

Baked good	Preparation and packaging	Freezer storage time*	To serve
CAKES			
Cheesecake	Prepare and bake as usual. Cool. Tray-freeze until firm; then wrap in freezer material. For added protection, store in a box or rigid container, sealing edges with freezer tape. Label.	4 months	Thaw in wrapping in refrigerator.
Fruitcake	Prepare and bake as usual. Cool. Wrap in freezer material, label, and freeze.	12 months	Thaw in wrapping at room temperature.
COOKIES			
	BAKED: Prepare and bake as usual. Cool. Package in rigid container or plastic freezer bag. Label.	6 to 8 months	Loosen wrapping and thaw at room temperature.
	UNBAKED: For *roll cookies*, form dough into rolls; wrap rolls in freezer paper; label. For *drop cookies*, drop dough onto waxed paper–lined cookie sheets; tray-freeze until firm. Package in rigid containers or plastic freezer bag; label.	6 months	For roll cookies, slice and bake without thawing. For drop cookies, bake without thawing.
PANCAKES AND WAFFLES			
	Prepare and cook as usual; cool. Stack, separating each layer with 2 sheets of freezer or waxed paper. Wrap stack in freezer material or store in plastic freezer bag. Label.	1 to 2 months	Heat without thawing in toaster, toaster-oven, hot oven, or under broiler.
PASTRY			
Cream puff and eclair shells	BAKED: Prepare and bake as usual. Cool. Tray-freeze until firm. Package in rigid containers or plastic freezer bags. Label.	1 to 2 months	Loosen wrapping and thaw at room temperature.
	UNBAKED: Prepare pastry and pipe or drop onto waxed paper-lined cookie sheets; tray-freeze until firm. Package in plastic freezer bags; label.	1 to 2 months	Remove wrapping and place on baking sheets; thaw and bake as usual. Or bake unthawed if desired.
Pie and tart shells	BAKED: Prepare and bake as usual. Cool. Freeze until firm. Stack, separating layers with 2 sheets of freezer or waxed paper. Wrap in foil or place in a plastic freezer bag. For additional protection, store in a box or rigid container, sealing edges with freezer tape. Label.	3 to 4 months	Loosen wrapping and thaw at room temperature. Unwrap and add filling.
	UNBAKED: Prepare pastry and line pie or tart pans. Freeze until firm. Wrap and package according to instructions for Pie and Tart Shells, Baked.	6 to 8 weeks	Bake in hot oven until browned. Or fill and bake according to recipe. Thawing is optional in either case.
Puff pastry	BAKED: Shape into shells and bake as usual. Cool. Tray-freeze until firm. Package in rigid containers or plastic freezer bags. Label.	1 to 2 months	Loosen wrapping and thaw at room temperature.
	UNBAKED: Prepare dough as usual. Wrap in bulk in freezer material; label.		Thaw in wrapping in refrigerator.

*Foods stored longer than the recommended period will suffer a loss of quality but will still be safe to eat.
Storing foods at temperatures higher than 0°F shortens the storage period considerably.

Baked good	Preparation and packaging	Freezer storage time*	To serve
Pies			
Fruit	UNBAKED: Prepare as usual, adding extra thickening: ½ to 1 tablespoon cornstarch or 1 to 2 tablespoons flour for each pie. Do not cut vents in top crust. Tray-freeze until firm. Wrap in aluminum foil and seal. For added protection, store in a box or rigid container, sealing edges with freezer tape. Label.	3 to 4 months	Thaw sufficiently to permit vents to be cut in top crust. Brush top with beaten egg or milk if desired. Bake according to recipe.
	BAKED: Prepare and bake as usual. Cool. Wrap and package according to instructions for Fruit Pies, Unbaked. Label.	6 months	Thaw in wrapping at room temperature or in low oven.
	FILLING: Prepare as usual, adding extra thickening. (See Fruit Pies, Unbaked.) Line pie pan with aluminum foil (see casserole-wrap method, page 13); pour in filling; freeze until firm. Remove from pan, seal foil, label, and return to freezer.	6 to 8 months	Peel off foil and place frozen filling in pastry-lined pan. Add top crust and bake according to recipe.
Mincemeat	Prepare and bake as usual. Cool. Wrap and package according to instructions for Fruit Pies, Unbaked.	6 to 8 months	Thaw in wrapping in refrigerator.
Nut	Prepare and bake as usual. Cool. Wrap and package according to instructions for Fruit Pies, Unbaked.	3 to 4 months	Thaw in wrapping in refrigerator.
Pumpkin	BAKED: Prepare and bake as usual. Cool. Wrap and package according to instructions for Fruit Pies, Unbaked.	4 months	Thaw in wrapping in refrigerator.
	UNBAKED: Prepare pie as usual. Tray-freeze until firm. Wrap and package according to instructions for Fruit Pies, Unbaked.	4 to 5 weeks	Bake without thawing.
Pizza			
	BAKED: Prepare and bake as usual, omitting sausage or pepperoni. Cool. Tray-freeze until firm. Wrap in aluminum foil; seal and label.	3 months	Loosen foil and bake with or without thawing in 375°F oven until heated through.
Sandwiches and Canapés			
Canapés	Cut bread into shapes. (Toast if desired.) Spread butter or margarine over entire surface. Add topping. (See page 48 for toppings that don't freeze well.) Tray-freeze until firm. Package in rigid container or plastic freezer bag, separating layers with freezer paper.	2 to 4 weeks	Arrange on serving trays and thaw at room temperature.
Sandwiches	Spread butter or margarine over entire slice of bread. Add choice of filling. (See page 48 for fillings that don't freeze well.) Top with buttered bread slice and wrap in freezer material; label.	3 to 4 weeks	Thaw in wrapping at room temperature.

*Foods stored longer than the recommended period will suffer a loss of quality but will still be safe to eat.
Storing foods at temperatures higher than 0°F shortens the storage period considerably.

BOSTON BROWN BREAD

Here is an American classic to savor with baked beans. Or spread slices with cream cheese for breakfast. Cook one loaf now and freeze the other for another occasion.

   Butter
   Cornmeal
  2 *cups buttermilk*
 3/4 *cup light molasses*
  1 *cup whole wheat flour*
 1/2 *cup all-purpose flour*
 1/2 *cup rye flour*
  1 *teaspoon salt*
 3/4 *teaspoon baking soda*
 1/2 *teaspoon baking powder*
  1 *cup cornmeal*
 3/4 *cup raisins*

1. Butter insides of two 1-pound coffee cans and dust with cornmeal.

2. In a large bowl, combine buttermilk and molasses; stir to blend.

3. Combine whole wheat, all-purpose, and rye flours with salt, baking soda, and baking powder. Add a third of the flour mixture to the buttermilk mixture, stirring just until combined. Repeat. Stir in cornmeal and raisins.

4. Divide batter evenly between cans. Cut 2 circles of waxed paper the diameter of the cans. Butter paper on one side and place, buttered side down, on batter. Cover tops of cans with heavy-duty foil, molding it to fit tightly over cans. Secure with string.*

5. Place cans in a large pot. Add enough boiling water to reach halfway up sides of cans. Cover pot with tight-fitting lid. Bring water to a gentle simmer and cook 2½ hours, or until tops spring back when pressed gently with fingers.

6. Run a knife along sides of cans and unmold bread carefully. Let cool 10 minutes before slicing. Serve hot.

Yield: 2 loaves

*To FREEZE BATTER: Seal foil to cans with freezer tape. Freeze up to 1 month. Remove freezer tape and cook as directed above. It's not necessary to thaw batter.

To FREEZE LOAVES: Cool completely. Wrap in heavy-duty foil and seal with freezer tape. To warm, remove tape and heat in 350°F oven for 15 to 20 minutes.

WHOLE WHEAT BREAD

  5 *cups all-purpose flour*
  2 *or 3 packages active dry yeast*
2¾ *cups water*
 1/2 *cup firmly packed brown sugar*
 1/4 *cup butter or margarine*
  1 *tablespoon salt*
  3 *cups whole wheat flour*
   Oil

1. In a large bowl, combine 3½ cups of the all-purpose flour and 2 packages of yeast. (Use 3 packages of yeast if you plan to freeze the dough.)

2. In a saucepan, heat water, brown sugar, butter, and salt just until warm (110°–115°F), stirring to melt butter. Add to dry ingredients. Beat batter with electric mixer about 4 minutes.

3. By hand, stir in whole wheat flour and enough remaining all-purpose flour to make a moderately stiff dough.

4. Turn onto a lightly floured surface and knead until smooth and elastic (10 to 12 minutes). Shape into a ball and place in a lightly greased bowl, turning to grease surface. Cover and let rise in a warm place until doubled (about 1 hour).

5. Punch dough down. Turn onto a lightly floured surface; divide in half. Cover and let rest 10 minutes.

6. Shape dough into 2 loaves. Place in two greased 8½- by 4½- by 2½-inch loaf pans.* Cover and let rise in a warm place until almost doubled (40 to 45 minutes).

7. Bake at 375°F for 40 to 45 minutes. If necessary, cover loaf with foil during last 20 minutes to prevent overbrowning.

Yield: 2 loaves

*To FREEZE UNBAKED DOUGH: Freeze in pans until firm. Remove from pans and wrap. Dough may be stored in freezer up to 3 weeks. To bake, remove wrap and return dough to greased loaf pans; cover. Thaw and let rise in warm place until doubled (2 to 4 hours). Bake as directed above.

To FREEZE BAKED LOAVES: Cool to room temperature and wrap. Freeze up to 6 months.

OREGON BLUEBERRY PIE

  Pastry for double-crust 9-inch pie
  2 *tablespoons melted unsalted butter or vegetable shortening*
 2/3 *cup sugar*
  2 *or 3 tablespoons cornstarch*
  4 *cups blueberries*
  1 *tablespoon lemon juice*
  1 *tablespoon butter or margarine*

1. Prepare pastry. Line pie pan with half. Brush bottom with melted butter.

2. Combine sugar and cornstarch. (Use 2 tablespoons of cornstarch if pie will be baked before freezing; 3 if pie will be frozen unbaked.) Toss with blueberries and lemon juice. If berries are tart, add sugar as needed. Fill shell with blueberries. Dot with butter.

3. Cover pie with top crust; seal and flute edges.*

4. Cut slits in top crust for steam vents. Bake at 425°F for 15 minutes. Reduce heat to 375°F and bake 25 to 35 minutes longer. If necessary, cover edge of crust with foil to prevent overbrowning.

Yield: One 9-inch pie

*To FREEZE BEFORE BAKING: Do not cut slits in top crust. Tray-freeze pie until firm. Wrap. Freeze up to 3 months. Bake unthawed, following Steps 4 and 5.

To FREEZE AFTER BAKING: Cool pie to room temperature. Wrap. Freeze up to 6 months.

BRAN-MOLASSES MUFFINS

Bran and molasses give this muffin a wholesome flavor. You can freeze the baked muffins or freeze the batter and bake just before serving. Frozen batter needs only 5 to 6 minutes' additional baking time.

- 1/4 cup butter or margarine, at room temperature
- 1/2 cup firmly packed brown sugar
- 1/4 cup light molasses
- 2 eggs
- 1 cup milk
- 1 1/2 cups bran cereal (buds rather than flakes)
- 1 cup all-purpose flour
- 1 1/2 teaspoons baking soda
- 3/4 teaspoon salt

1. In a large bowl, cream butter and brown sugar. Add molasses and eggs and beat well.
2. Mix in milk, and then bran.
3. Combine flour, baking soda, and salt. Stir into bran mixture until blended. Do not overmix.
4. Divide batter among paper-lined muffin pans, filling each cup about two-thirds full.*
5. Bake at 400°F for 15 minutes. Serve warm.

Yield: About 15 muffins

*TO FREEZE BATTER: Freeze in pans until firm. Remove from pans and wrap. To bake, return frozen muffins to pans. Bake as directed above, adding 5 to 6 minutes to baking time.

TO FREEZE BAKED MUF-FINS: Cool to room temperature and wrap. Freeze up to 6 months.

OLD-FASHIONED BUTTERMILK CAKE

Fill this beautiful triple-layer cake with whipped cream and berries.

- 1 cup butter
- 2 cups sugar
- 4 eggs
- 3 cups sifted cake flour
- 1 tablespoon baking powder
- 1/2 teaspoon baking soda
- 1/2 teaspoon salt
- 1 1/4 cups buttermilk
- 1 teaspoon vanilla

1. Grease and flour three 9-inch round cake pans.
2. In a large mixing bowl, cream butter and sugar until fluffy. Add eggs one at a time, beating after each addition.
3. Combine flour, baking powder, baking soda, and salt. Mix into creamed mixture alternately with buttermilk, beating after each addition until batter is smooth. Add vanilla with last of buttermilk.
4. Divide batter among pans.*
5. Bake at 350°F for 25 to 35 minutes, or until top springs back when pressed lightly. Cool on wire racks.

Yield: 3 cake layers

*TO FREEZE BATTER: Wrap pans and freeze. Or line pans with heavy-duty aluminum foil before filling; add batter and freeze. Remove from pans and wrap. Freeze up to 3 weeks. To bake, peel away foil and place frozen batter in greased and floured pans. Bake as directed above.

TO FREEZE BAKED LAYERS: Cool to room temperature and wrap. Freeze up to 6 months.

SOUR CREAM COFFEE CAKE

Expecting guests for Sunday brunch? Make this coffee cake and freeze the batter. All the work can be done up to three weeks before the party.

- 1 cup butter or margarine
- 1 1/4 cups sugar
- 2 eggs, lightly beaten
- 1 cup sour cream
- 1 teaspoon vanilla
- 2 cups sifted all-purpose flour
- 1 teaspoon baking powder
- 1/2 teaspoon baking soda
- Topping (recipe follows)
- Powdered sugar (optional)

1. In a large mixing bowl, cream butter and sugar. Beat in eggs, sour cream, and vanilla.
2. Combine flour, baking powder, and baking soda. Gradually beat into batter.
3. Butter and flour a 9-inch bundt or angel cake pan. Sprinkle in one-fourth of the topping mixture. Spoon on one-third of the batter. Continue layering, ending with topping mixture.*
4. Bake at 350°F for about 1 hour, or until a pick inserted in center comes out clean. Cool 15 minutes before removing from pan. Sprinkle top with powdered sugar if desired.

Yield: One 9-inch bundt coffee cake

TOPPING: Combine 1 cup chopped nuts, 1/4 cup sugar, and 1 teaspoon ground cinnamon.

*TO FREEZE BATTER: Wrap pan and freeze. Batter may be frozen up to 3 weeks. To bake, remove wrap and bake as directed above. Additional baking time is not required.

TO FREEZE BAKED CAKE: Cool to room temperature and wrap. Freeze up to 8 months.

LOCKWOOD BUTTER COOKIES

Be sure to prepare a big batch of these melt-in-your-mouth butter cookies. Roll them in colored sugar crystals and you are ready for any occasion.

- 1 pound butter, softened
- 1 cup sugar
- 4 1/2 cups all-purpose flour
- 1 teaspoon vanilla
- Sugar crystals

1. In a large bowl, cream butter and sugar. Mix in flour and vanilla.
2. Shape into two rolls, each 1 1/2 to 2 inches in diameter. Roll in sugar crystals.*
3. Wrap in plastic freezer wrap or aluminum foil and freeze until firm.
4. Remove from freezer and let soften slightly. Cut in thin slices. Bake on ungreased cookie sheets at 300°F for about 10 minutes, or until bottoms are barely golden. Let cool on baking pan.

Yield: 6 1/2 dozen 2-inch cookies

*TO FREEZE DOUGH: Wrap rolls in plastic freezer wrap or aluminum foil and freeze up to 6 months. Slice and bake as directed above.

TO FREEZE BAKED COOKIES: Cool to room temperature. Package in rigid containers or freezer wrapping material. Freeze up to 8 months.

Bake ahead and freeze Bran Muffins, Boston Brown Bread, Lockwood Butter Cookies, Whole Wheat Bread, and Sour Cream Coffee Cake.

Tart lemon ice is a refreshing finale to any meal.

Freezing Eggs & Dairy Products

Although the majority of your freezer space is likely to be devoted to fruits, vegetables, meats, and baked goods, the freezer is also the ideal place to house a half dozen egg whites that you can't use immediately or two dozen eggs when the supermarket has them on special. A freezer also makes it possible to extend the storage life of cheese, stock up on butter when it's an exceptionally good buy, save a cup of leftover whipped cream from Sunday's strawberry short-cake, or savor the just-made flavor of homemade ice cream weeks after it's been churned.

Freezing is virtually the only safe method of home-preserving eggs and most dairy products.

The freezer is a great place to store leftover bits of cheeses like Cheddar, Jack, and blue.

You'll find detailed instructions for freezing eggs, whole and sepa-rated, and all kinds of dairy prod-ucts in this chapter. It also tells you how long these items can be stored safely.

And there are recipes for stocking your freezer with make-ahead treats—from a savory quiche to luscious frozen dessert confections like Fresh Lemon Ice Cream, Almond Java Ice Cream Pie, and Frozen Strawberry Soufflé.

Thaw eggs in their cov-
ered container in the
refrigerator. Transfer
cubes to a bowl and
cover with plastic film.
A pint of eggs takes 8 to
10 hours to thaw in the
refrigerator—all the
more reason to package
eggs in small amounts.
Eggs may also be
thawed in their con-
tainers under cold
running water. Never
refreeze thawed eggs.

How to Freeze Eggs

Since they expand during freezing, raw eggs cannot be frozen in the shell. And you won't want to freeze hard-cooked eggs because the whites become rubbery. But, removed from their shells, whole or separated raw eggs can safely be stored in the freezer for 9 to 12 months.

When freezing eggs, start with the very freshest. Make sure the shells are un-cracked, and wash soiled eggs thoroughly and wipe dry. After cracking, handle carefully to avoid bacterial growth. A really fresh egg has a firm white and a plump, high-standing yolk.

Freeze eggs in meal-, serving-, or recipe-size portions that can be used at one time. Small amounts also thaw more quickly. Ice cube trays or muffin pans make ideal containers for tray-freezing whole or separated eggs; when firm, place individual cubes in plastic freezer bags. Calculate the capacity of a single ice cube mold or muffin cup: How many whites or yolks will it hold? Can it accommodate a whole egg?

Whole eggs. To prepare whole eggs for freezing, stir gently to blend the yolk and white, being careful not to build up foam —air bubbles beaten into the egg will cause it to dry out during freezing. To prevent the yolk from thickening and becoming gummy during freezing, to each cup of egg add ½ teaspoon salt for even-tual main dishes or 1 tablespoon sugar for desserts. Package in plastic containers, allowing ½-inch headspace for expan-sion. Or tray-freeze unwrapped in ice cube trays or muffin pans. When cubes are firm, remove from trays or pans and package in plastic freezer bags.

Yolks and whites. Egg yolks and whites are frozen in a similar fashion to whole eggs. Separate the eggs and stir the yolks slightly, without foaming. Add ½ teaspoon salt or 1½ teaspoons sugar to each cup of egg yolk before freezing. Freeze egg whites without stirring or adding salt or sugar. Frozen egg whites whip just as well after thawing as when they are "fresh." Package yolks or whites according to the instructions for whole eggs.

Quiche any Style

Savory quiches are fabulous for brunch, supper, and picnics or as appetizers. They freeze successfully up to 2 months after baking or for 1 month before baking. Use any favorite filling: bacon and green chiles, onion and cheese, or ham and artichoke hearts. See the recipe for Tomato and Bacon Quiche opposite.

To freeze quiche before baking.
Tray-freeze until firm; then wrap the dish with freezer wrapping material. Label and freeze up to 1 month. Unwrap and bake as usual, allowing 10 to 20 minutes' additional time.

To freeze quiche after baking.
Cool to room temperature. Wrap, label, and freeze up to 2 months. To serve, do not thaw; rather, unwrap quiche and heat in a 350°F oven for 20 to 25 minutes, or until heated through.

Equivalents for Large Fresh Eggs

1 egg yolk	1 tablespoon stirred egg yolk
1 egg white	2 tablespoons egg white
1 whole egg	3 tablespoons stirred whole egg
5 whole eggs	1 cup stirred whole egg
10 whole eggs	1 pint stirred whole egg
16 whites	1 pint egg white
24 yolks	1 pint stirred egg yolk

TOMATO AND BACON QUICHE

A versatile dish, quiche is appropriate for any number of occasions. Prepare several at one time to store in the freezer and serve when a savory, light dish would be just right.

A 9-inch pastry shell

6 slices bacon, cut in 1-inch pieces

1/2 cup chopped onion

2 small tomatoes

3 eggs

1 1/2 cups half-and-half (light cream)

1/4 teaspoon dried basil

1/4 teaspoon dried oregano

1/2 teaspoon salt

1/8 teaspoon pepper

1/2 cup shredded Gruyère or Emmenthaler cheese

1 can (2 1/2 oz) sliced black olives

1. Partially bake (blind-bake) pastry shell.

2. Fry bacon in skillet. Drain on paper towels. Pour off all but 1 tablespoon fat from pan. Add onions and sauté until soft and transparent.

3. Peel tomatoes and slice thinly. Gently rinse slices under running water to remove seeds and pat slices dry.

4. Whisk eggs in a large bowl. Beat in half-and-half, basil, oregano, salt, pepper, and cheese.

5. Spread bacon and onion evenly over bottom of pastry shell. Place tomato slices on top. Sprinkle with sliced olives. Pour in egg mixture.*

6. Bake at 375°F for 25 to 30 minutes or until quiche puffs and is a light golden brown.

Yield: One 9-inch quiche

*TO FREEZE BEFORE BAKING: See instructions opposite.

TO FREEZE BAKED QUICHE: See instructions opposite.

You can freeze Tomato and Bacon Quiche before or after you bake it.

For convenient individual servings and faster thawing, shape butter into curls, balls, or pats before freezing. Mix butter with herbs or other flavorings before shaping, for festive party servings.

HOW TO FREEZE DAIRY PRODUCTS

Nearly every item in the supermarket dairy case—with the exception of half-and-half (light cream), sour cream, and cottage cheese—can be frozen. Although they separate when thawed, buttermilk and yogurt can be frozen for use in baking.

Homogenized milk, including 2 percent (low-fat) and nonfat milk, can be frozen as purchased in its sealed carton or transferred to freezer containers and sealed, allowing headspace for expansion.

Heavy (whipping) cream can be frozen unwhipped. This is a great way to save leftover cream for cereal or desserts. However, keep in mind that heavy cream that has been frozen will not whip to the usual volume. For best results, whip cream before freezing, adding sugar to taste. Tray-freeze in mounds and store in plastic freezer bags. Thawing takes only

minutes. Try this with extra whipped cream from strawberry shortcake.

Butter and margarine can go directly from store to freezer in their original cartons, overwrapped in freezer material or in plastic freezer bags. Unsalted butter loses its flavor faster than salted butter, so its storage time is shorter. Flavored butters hold up well in the freezer. Herb butters (see pages 32 and 33) are delicious make-ahead-and-freeze spreads for breads or flavorings for vegetables and grilled meats. Thaw butter in the refrigerator or at room temperature. Butter may be refrozen.

Cheeses that freeze well are the semihard and hard cheeses, soft cheeses, and cream cheese. Semihard and hard cheeses include Parmesan, Cheddar, Havarti, Münster, Gruyère, Swiss, Gouda, Edam, Monterey jack, mozzarella, and the blue-veined cheeses; Camembert and Brie are examples of soft cheeses.

Semihard and hard cheese can be frozen sliced, grated, or cut into blocks of convenient size. Frozen hard cheeses may be grated and then refrozen. Wrap blocks of cheese in freezer material; separate slices with a double thickness of waxed or freezer paper; store grated cheese in plastic freezer bags or containers. Hard or semihard cheeses may become crumbly when thawed, but their flavor will not change. They are perfect in dishes that call for grated cheese or those in which the cheese is melted.

Soft cheeses should be frozen when they have reached the desired degree of ripeness; wrap in freezer material. Overwrap prewrapped bricks of cream cheese before freezing.

Always thaw cheese in the refrigerator; thawing at room temperature causes cheese to crumble.

Ice cream, ice milk, sherbets, and ices should be stored in plastic freezer containers with tight-fitting lids. To maximize the storage time of commercial ice cream, place its container in a plastic freezer bag. To help keep ice crystals from forming after you scoop ice cream, lay clear plastic wrap directly on the scooped-out surface. Homemade ice cream pies, cakes, and rolls should be tray-frozen and then wrapped in aluminum foil or other freezer material. For added protection, store in a box, sealing the edges with freezer tape.

TIMETABLE FOR STORING EGGS AND DAIRY PRODUCTS

	Storage time at 0°F*
Butter	
Salted	6 months
Unsalted	2 to 3 months
Buttermilk, for baking purposes	1 month
Cheese	
Cream cheese	2 months
Hard and semihard natural cheese	2 to 3 months
Pasteurized process cheese	4 months
Soft cheese	2 months
Cream	
Heavy (whipping)	2 months
Whipped	1 month
Eggs: whole, yolks, or whites	9 to 12 months
Ice cream, ice milk, sherbets, and ices	1 to 2 months
Margarine	12 months
Milk	3 months
Yogurt, for baking purposes	1 month

*Foods stored longer than the recommended period will suffer a loss of quality but will be safe to eat. Storing foods at temperatures higher than 0°F shortens the storage period considerably.

HAVE YOU THOUGHT TO FREEZE...

Bread crumbs and croutons. Don't throw away stale bread. Turn it into crumbs in the blender or food processor or cut it into cubes for croutons; then freeze.

Brown sugar. Overwrap store boxes or packages in plastic freezer bags; sugar will be soft when thawed.

Candied or glacé fruits. Stock up at after-holiday sales and freeze for next season's fruitcake. Overwrapped in plastic freezer bags, fruits keep a year or longer.

Candy. Divinity, fudge, toffee, caramel corn, and creams—homemade or commercially prepared—can be frozen for up to a year. Wrap tightly in freezer material or in plastic freezer bags.

Cereals and grains. A temperature of 0°F kills insects and their larvae in 2 to 30 days; if you don't have adequate space to store cereals and grains in the freezer permanently, "debug" them briefly after purchase.

Citrus peel. When you squeeze a lemon, orange, or grapefruit, freeze the peel to grate later, grate it before freezing, or slice it into twists and freeze to add to drinks. Use grated peel to flavor baked goods, frostings, and butter sauces.

Coffee. Keep ground or whole beans fresh in the freezer. If you purchase coffee in brown paper bags rather than cans, overwrap before freezing.

Crackers and chips. These stay crisp when packaged tightly and frozen.

Dried fruits. Prunes, raisins, dates, and apricots stay fresh and moist for at least a year in the freezer, and they're easier to chop when frozen.

Homemade baby food. Process in quantity and freeze in tiny meal-size portions. Ice cube trays, paper cups, and muffin tins make handy preportioned freezing compartments. Tray-freeze (see page 14) and then store cubes in plastic freezer bags. Prepare and package baby food carefully, using the very freshest foods and scrupulously clean utensils, equipment, and containers. Follow storage times for specific types of food and thaw in the refrigerator.

Marshmallows. Frozen, they're easy to cut and won't stick to knife or fingers.

Nuts. Although they're especially prone to rancidity and insect infestation at room temperature, nuts keep well in the freezer for at least 6 months. (See "Cereals and grains" for information on "debugging.") Salted or seasoned nuts keep only about half to two-thirds as long as unsalted, unflavored nuts.

Olive and other oils. To prevent rancidity, divide large quantities into smaller containers, leaving headspace for expansion (¼ inch for pints; ½ inch for quarts), and freeze. You can also strain and freeze oil that has been used in deep-fat frying for reuse. Oil clouds when frozen, but clears when it thaws.

Party supplies and nibbles. Freeze pickles, pimientos, olives, water chestnuts, and bamboo shoots in their own brine or liquid. Keep in mind, however, that their texture will suffer somewhat. See page 48 for freezing hors d'oeuvres.

Sauces. Frozen white, brown, cheese, curry, and tomato sauces in containers or ice-cube tray portions add spice to dinner entrées. Cheese and curry sauces keep up to 3 months; white and brown sauces, 4 to 6 months; and tomato sauce, 1 year.

Wine. Freeze leftover amounts in small containers or ice cube trays to flavor casseroles, soups, sauces, and stews.

Woolens. Package in heavy plastic bags and seal well. Freeze for at least 72 hours to destroy moths.

ODDS 'N ENDS

Ice cube trays and muffin tins make handy containers for small amounts. Calculate the capacity of an individual cup so that you know how much it holds. Tray-freeze (see page 14) foods until firm; then remove from tray or tin and package cubes in plastic freezer bags. Try this approach for fruit purées and applesauce, leftover tomato paste or sauce, homemade baby food, egg whites or whole eggs, citrus juices, stocks, leftover tea, and party ice cubes. (Add food coloring to water for colored cubes or drop a maraschino cherry or a grape into each cup.)

Use your home freezer to make ice creams, ices, and sherbets like this dazzling raspberry one, for refreshing dessert treats.

How to Freeze Dairy Desserts

Desserts that have a cooked, gelatin, or whipped-cream base—like un-baked soufflés, bavarians, and mousses—freeze well. Those with a whipped-egg-white base do not. And of course, ice cream desserts like rolls and pies are naturals for freezing. Mix mousses and pour into rigid freezer containers. Freeze bavarians before they set; they will set up as they thaw. Both keep 2 months, and ice cream desserts keep for 6 to 8 weeks. All should be thawed in the refrigerator until soft enough to serve. Try one of the luscious frozen dessert recipes at right.

ALMOND JAVA ICE CREAM PIE

This recipe pairs coffee and chocolate-almond ice cream in a dessert everyone will love. Experiment with different ice cream combinations to create your own rainbows.

1/4 cup semisweet chocolate chips, finely chopped

1 cup graham cracker crumbs

4 tablespoons butter or margarine, melted

1 1/2 cups (12 oz) coffee ice cream

2 tablespoons brandy

1 1/2 cups (12 oz) chocolate-almond or chocolate ice cream

Hot fudge topping (optional)

Sliced toasted almonds

1. In a medium bowl, combine chopped chocolate chips, graham cracker crumbs, and melted butter. Firmly press against bottom and sides of a 9- or 10-inch springform pan. Bake at 300°F for 10 minutes or until set. Cool on wire rack.

2. Beat coffee ice cream and brandy until smooth. Pour into cooled crust. Freeze until firm.

3. Beat chocolate-almond ice cream until smooth. Pour over coffee ice cream

layer and spread evenly in pan. Wrap pie in freezer wrapping material. Freeze up to 6 weeks.

4. To serve, remove pie from springform pan and transfer to serving dish. If desired, heat fudge topping and spoon over pie. Garnish top with almonds.

Yield: One 9- or 10-inch pie (8 to 12 servings)

FRESH LEMON ICE CREAM

For an elegant presentation, serve this ice cream in scooped-out lemon shells.

- 2 cups whipping cream
- 1 cup sugar
- 1 tablespoon finely grated lemon peel
- 1/3 cup fresh lemon juice
- 8 to 10 lemon shells or boats (optional)
  Fresh mint for garnish (optional)

1. In a large bowl, stir together cream and sugar until sugar dissolves. Blend in lemon peel and lemon juice.

2. Pour mixture into a shallow pan. Freeze until firm (about 4 hours).

3. Serve in lemon shells or boats or in dessert bowls, garnished with fresh mint leaves if desired. Or spoon into rigid freezer containers and store up to 2 months.

Yield: 3 cups

ORANGE CREAM SHERBET

When you make sherbets and ice cream in the freezer, the more you beat the mixture during the freezing period, the smoother its texture.

- 2 cups half-and-half (light cream)
- 3 cups fresh orange juice
- 3 tablespoons orange liqueur
- 2 tablespoons lemon juice
- 1 to 1 1/3 cups sugar

1. Combine half-and-half, orange juice, orange liqueur, and lemon juice. Add sugar and stir until it dissolves.

2. Freeze in ice cream maker following manufacturer's instructions. Or pour into shallow pans and freeze until sherbet is firm about 1 inch around sides of pan; remove from freezer and beat vigorously until smooth; return to freezer. Repeat beating and freezing process 2 or 3 times during the freezing period (2 to 3 hours).

3. Serve immediately or spoon into rigid freezer containers and store up to 2 months.

Yield: About 1 1/2 quarts

STRAWBERRY SOUFFLÉ

Frozen soufflés are as delicious as they are elegant. This soufflé may be frozen in individual soufflé dishes as well.

- 6 large eggs, separated
- 1 3/4 cups sugar
- 2 cups puréed fresh strawberries
- 1/3 to 1/2 cup orange liqueur
- 1/3 cup fresh orange juice
- 3 cups whipping cream
  Whole strawberries for garnish

1. Prepare 1 1/2-quart soufflé dish with a lightly oiled 2 1/2-inch collar of heavy-duty foil.

2. In a large bowl, beat egg yolks with 3/4 cup of the sugar until thick and lemon colored. Stir in 1 cup of the puréed strawberries.

3. Cook and stir yolk mixture in top of double boiler until thickened. Do not boil. Remove from heat and cool completely. Blend in orange liqueur.

4. In a separate saucepan, combine remaining 1 cup sugar with orange juice. Cook over medium-low heat, stirring to dissolve sugar. Continue cooking without stirring until liquid reaches soft-ball stage (245°F).

5. While sugar and orange juice cook, beat egg whites until soft peaks form. Slowly pour in hot orange syrup, beating until stiff peaks form.

6. Whip cream until soft peaks form. Fold into cooled yolk mixture. Fold in remaining strawberry purée. Gently fold in egg white mixture. Spoon into soufflé dish. Freeze until firm (about 2 to 3 hours) or wrap in freezer material and freeze up to 2 days.

7. To serve, remove freezer wrap (if used) and collar. Decorate top with whole strawberries.

Yield: 12 to 16 servings

FROZEN HONEY YOGURT

Light and tangy, frozen yogurt is a refreshing summer treat. It's especially good topped with fresh fruit or a lightly sweetened fruit purée.

- 3 cups plain yogurt
- 1/4 cup honey
- 1 cup whipping cream
- 1/4 cup firmly packed brown sugar
- 1 teaspoon almond extract

1. In a large bowl, mix yogurt and honey until smooth. Place in freezer while completing Step 2.

2. In a separate bowl, whip cream with brown sugar and almond extract until soft peaks form. Remove yogurt mixture from freezer; fold whipped cream into yogurt.

3. Pour mixture into a shallow pan. Freeze until firm about 1 inch around sides of pan; remove from freezer and beat vigorously until smooth; return to freezer. Repeat beating and freezing process 2 or 3 times during the freezing period (2 to 3 hours).

4. Serve immediately or spoon into rigid freezer containers and store up to 2 months.

Yield: 1 quart

Rings of dried pineapple make chewy, sweet snacks.

Drying & Smoke-Cooking

Whether you enjoy them as a snack, in a trail mix, or as an ingredient in baking or cooking, dried fruits add a new dimension to home-preserving. Raisins, dates, and apricots are familiar staples in most kitchens, but the chewy sweetness of home-dried peaches, apples, pineapple, and figs may come as a welcome surprise.

You'll dry vegetables mainly for convenience—to make hearty quick soups, to use as flavorings, and for economical camp meals. If you grow herbs, you'll certainly want to preserve their heady flavors to spark winter menus. While they're drying, herbs and flavorings also make unusual decorations: See the instructions for chile-pepper and citrus-peel garlands.

This chapter also tells you how to dry flowers for potpourri or to decorate pastries and how to make your own jerky for snacks. And for barbecue fans, there's a section on preparing moist, succulent, flavorful smoked meats in a water smoker or a covered barbecue.

Drying is a relatively inexpensive way to preserve food. You pay only for the energy to heat the dehydrator or oven, and drying in the sun is free. Since dried food is only a fraction of its original size, you also save storage space.

Colorful dried vegetables should be packaged in airtight containers and stored in a cool, dry, dark place.

Drying is one of the most practical methods of food preservation. The only piece of equipment you may want is a dehydrator; you can also dry foods in the oven or in the sun.

METHODS OF DRYING

The key to drying food successfully is controlling both temperature and air circulation around it. Warmth causes excess moisture in the food to evaporate; movement of air over the food carries the moisture away. (Home-dried fruits and vegetables retain only 5 to 20 percent of their original moisture.) If the temperature is too low, food will dry too slowly and may spoil. If the temperature is too high, food will cook, or its outside surface will harden, locking moisture inside. This is called case-hardening, and it will cause the food to deteriorate eventually.

There are several methods of home-drying: in a dehydrator, in an oven, or in the sun. This chapter describes these methods as well as specific guidelines for drying in a microwave oven, a convection oven, and at room temperature. The method you choose will depend on the amount of time and money you wish to invest in drying and the types and quantities of food you intend to dry.

HISTORICAL NOTES

The ancient Egyptians and Greeks are known to have dried foods, and as early as 1490 B.C., sun-dried grapes (raisins) were important in the nomadic commerce of the ancient Middle East. Early explorers of this continent relied on dried fruits and meats for survival during months of sea travel. When the first colonists arrived in the New World, its inhabitants were already using fire and sun to dry much of their food. In fact, the Indians are said to have taught the early settlers how to dry corn and grind it into meal.

Commercial drying began in France in 1795, with the invention of the first dehydrator, but not until World War I was there a real demand for dried food—to feed the thousands of fighting troops. During World War II, more than 150 dehydration plants began operating in the United States to meet the needs of the armed services.

PRETREATING FOODS BEFORE DRYING

To help preserve their color and to stabilize them during storage, most fruits and vegetables should be pretreated before drying. Fruits that darken when cut and exposed to air should be pretreated with an antioxidant solution or a honey-water dip (see pages 72 and 75) before dehydrator- or oven-drying. If you sun-dry fruits, sulfur those that darken (see pages 74 and 75) or dip them in an antioxidant or a honey-water solution. (It's not necessary to do both.) The chart on page 76 indicates which fruits darken when exposed to air.

Most vegetables should be pretreated by blanching before drying. See page 28 for blanching instructions and the chart on pages 82 and 83 for specific vegetables that should be blanched.

DRYING IN A DEHYDRATOR

Using a dehydrator is by far the simplest method of drying because it involves fewer variables than sun- or oven-drying. You can dry virtually any food in a dehydrator, and on the whole, dehydrator-dried food has a better appearance than food dried in the sun or an oven. Food also dries more evenly in a dehydrator and it's almost impossible to scorch or overdry food. The appliance can be left unattended and can be operated day and night. Finally, dehydrator-drying—unlike sun-drying—is not dependent on climatic conditions. Dehydrators perform best indoors, in a dry, well-ventilated room. Operating instructions vary from model to model, so consult your manufacturer's manual.

Ready-made and homemade dehydrators. Dehydrators contain heating elements that draw moisture from the food and a fan that blows warm air across the food to absorb the released moisture and carry it away. You can choose a ready-made dehydrator or construct your own.

You can purchase manufactured dehydrators at department stores, hardware stores, or through mail-order catalogs. Prices range from $50 to $300 or more, depending on size, construction, and design. If you plan to dry large quantities of food, an electric factory-made dehydrator is a wise investment.

Dehydrators are inexpensive to build. For directions on how to make your own dehydrator, write for Circular 855, *How to Build a Portable Electric Food Dehydrator,* by Dale E. Kirk, Agricultural Engineer, Oregon State University, Corvallis, OR 97331.

Size. Dehydrators are available in a variety of sizes. The *outside* dimensions of the dehydrator tell you how much space you'll need to store the appliance. The number and size of trays *inside* the dehydrator tell you the amount of drying space available. A dehydrator containing ten 12- by 12-inch trays, for example, has 10 square feet of drying area. Generally, each square foot of tray space will dry about 1½ to 2½ pounds of prepared food.

Heat source. An efficient heat source will dry food in a short enough time to prevent spoilage but not so fast that the food cooks instead of dries. Manufactured dehydrators have rod-type heating elements. Homemade dehydrators generally use light bulbs or porcelain cones. The most efficient location for the heating rod is at the side or back of the dehydrator because the air can flow evenly across the food. Heating elements that are positioned on the bottom may overcook food on lower trays; to prevent this, trays will need to be rotated often.

Temperature control. A thermostat that controls the dehydrator's temperature is essential. The thermostat should be adjustable, or it will be difficult to dry each food at its ideal temperature. For example, herbs dry at 100°F or lower; fruits, vegetables, and meats dry at about 120° to 140°F.

Air circulation. Look for dehydrators with a fan or blower that produces a high-velocity air flow. This is particularly important if you plan to dry high-moisture foods like peaches and pears.

The best-constructed dehydrators have air vents on the left and right sides or in the front and back. Vents allow moist air to escape and fresh air to enter. Models without vents rely on the door for air exhaust and intake.

Insulation. Insulation greatly reduces heat loss, improving the energy efficiency of the appliance. Check the door for a snug fit and see that it latches securely.

Trays. Dehydrator trays should be a fine mesh, for maximum air flow, and the mesh should be small enough so that food will not fall through. Make certain the trays slide in and out of the dehydrator easily. For proper air circulation, trays should

be spaced at least 1¼ inches apart. For more information on tray construction, see page 69.

How to dehydrator-dry. Before beginning a drying project, read the information later in this chapter on recommended pretreatment, temperatures, drying times, and dryness indicators for specific foods. *Sulfur-treated foods (see page 74) should never be dried in a dehydrator.*

To begin drying, spread food on drying trays in a single layer; pieces should not touch each other. Different foods may be dried at the same time, but foods with strong odors—like fish, onion, and garlic—should be dried separately. Some dehydrator manufacturers recommend that you rotate the trays, front to back and top to bottom, at least once during the drying period. Turn the pieces of food once or twice, but wait to turn until any juice in the cavities of fruit like peaches and apricots has disappeared. Toward the end of the recommended drying period, check the food for dryness. (See page 68.) When food has dried, cool and store it following the instructions on page 70.

You can dry herbs, flavorings, and flowers successfully indoors on small trays at room temperature. A stacking arrangement like this one is convenient. See pages 84-89 for drying instructions. For the best results with fruits and vegetables, choose a dehydrator, your oven, or the sun.

TESTING FOODS FOR DRYNESS

The easiest way to determine whether food is dry is to touch and taste it. For accurate assessment, the test samples must be cool. Foods still warm from the sun, oven, or dehydrator will seem softer, more pliable, and moister than they actually are.

In general, dried fruit should be chewy and leatherlike, with no pockets of moisture. Cut a piece of fruit and press—no moisture should drip from the fruit. Or squeeze several pieces together—they should fall apart when released. Overdried fruits will have a strawlike texture and will have lost much of their nutritive value, color, and flavor. Guidelines for determining whether specific fruits are dry are provided in the chart on page 76.

Dried vegetables should generally have a brittle or tough-to-brittle consistency. Consult the chart on pages 82 and 83 for dryness indicators for specific vegetables.

DRYING IN AN OVEN

Oven-drying involves little or no investment in equipment. It is especially suited for drying a small quantity of food and in areas where sun-drying isn't possible.

The oven-drying process is similar to dehydrator-drying—trays of food are heated at low temperatures over a period of time. However, it differs from dehydrator-drying in several ways. It may take longer and will use more energy than dehydrator-drying. Since the oven door must remain open during the drying process to allow excess moisture to escape, the process adds heat to the house. And it ties up the oven for hours, so you'll need to keep this in mind when selecting a time to dry.

Whether your oven is suitable for drying depends on the *lowest* temperature it can achieve. Test your oven by heating it to its lowest setting, opening the door 4 to 6 inches, and measuring the temperature with an oven thermometer. The oven should maintain a temperature of 120° to 140°F. Too high a temperature will cook food rather than dry it. Unfortunately, some older ovens offer a minimum temperature too high for oven-drying.

Heat should come only from the bottom of the oven. If your oven has a top heating element that gives off heat when the "bake" setting is on, remove the top element or place a large cookie sheet on the top shelf as close to the heating unit as possible to deflect heat.

Options for trays to use in oven-drying are described on the opposite page. Baking sheets are not recommended because they do not allow air to circulate all around the food. Tray size depends on the size of your oven. To allow air movement, trays should be at least 1 inch smaller than the inside of the oven. Use blocks or bricks to stack trays about 3 inches apart, allowing at least 3 inches of clearance at the bottom and top of the oven.

How to oven-dry. Before beginning a drying project, read the information later in this chapter on recommended pre-treatments, temperatures, drying times, and dryness indicators for specific foods.

Spread food on trays in a single layer; pieces should not touch each other. Foods with strong odors—like garlic, onion, and fish—should be dried separately. (Keep in mind that in oven-drying their odors will permeate the house.) Preheat the oven to 140°F and add the loaded trays. Place a thermometer on the top tray toward the back of the oven and monitor the temperature frequently, adjusting the oven thermostat to maintain a temperature between 120° and 140°F. To improve air circulation, prop the oven door open 4 to 6 inches and use an electric fan to keep moist air from accumulating. Place the fan on a chair or stool outside the oven, position it to move air through the opening and across the oven, and set the fan on "low." Change the position of the fan from one side of the oven to the other every few hours during drying.

Rotate the trays from top to bottom and front to back every 2 to 3 hours. Turn pieces of food at least once. Food at the outer edges of the trays will dry faster, so remove it as it becomes dry. Watch food carefully toward the end of the drying period because oven-dried foods can scorch easily when they are almost dry.

Note: If you need the oven for cooking before food is dry, remove the trays and place in a dry spot. When you have finished cooking, allow the oven to cool to 120° to 140°F, replace the trays, and continue drying. The more frequently you remove the trays from the oven, the longer it will take the food to dry. Remember that food reabsorbs moisture at room temperature.

Drying in a microwave oven. Microwave ovens make it possible to dry fresh herbs or grated citrus peel in a matter of minutes. See pages 84 and 86 for instructions. Microwave ovens are not suitable for drying other foods.

Drying in a convection oven. Convection ovens cook food with fan-driven, electrically heated air that is circulated evenly to all food surfaces. This circulating-air feature makes a convection oven an ideal chamber for drying fruits, vegetables, and meats. Many models offer optional dehydration racks specifically designed to dry foods. Consult your manufacturer's manual for detailed instructions on convection oven–drying.

DRYING IN THE SUN

Sun-drying is by far the oldest method of preserving food. Although it takes longer than other methods—4 to 5 days of hot sun for most foods—it requires little investment. It's an ideal method in areas that have consecutive hot, dry days with temperatures in the nineties or higher, low humidity, and relatively clean air—for example, central California, the Southwest, and parts of the Midwest. Avoid sun-drying in areas where the air is dusty or animals are housed and near heavily traveled roads. People living in humid areas like the South have had unsatisfactory results when sun-drying.

Certain fruits and vegetables are recommended for sun-drying because they won't spoil during the several-day drying period: apples, apricots, cherries, figs, grapes, nectarines, peaches, pears, plums, citrus peels, and pineapple among the fruits and chile peppers, lentils, peas, shell beans, and sweet corn among the vegetables. Meats and fish should not be sun-dried because there is a risk of spoilage.

The only limit to the amount of food you can sun-dry at one time is the number of trays you have on hand. Types of trays to use in sun-drying are discussed at right.

How to sun-dry. Before beginning a drying project, read the information later in this chapter on recommended pretreatment, temperatures, drying times, and dryness indicators for specific foods.

Spread food on trays in a single layer; pieces should not touch each other. Place trays flat in direct sunlight on a platform,

TRAYS FOR DRYING

Drying trays can be simple cooling racks from your kitchen or elaborate home-constructed wooden models. Ultimately, your choice will depend on the drying method you select, the amount of time and money you want to invest, and the materials or equipment you have on hand.

Factory-made dehydrators usually come with their own trays. Most are made of plastic-coated or stainless steel screen embedded in a plastic or metal frame. Trays made entirely of plastic may warp and sag over time. Dehydrator trays are easy to maintain and clean—simply soak for a few minutes in warm water and wipe clean.

Trays for oven-drying and sun-drying are interchangeable. If you're planning to dry only a small quantity of food, rely on readily available tray materials. Cooling racks tightly covered with several layers of cheesecloth or nylon netting (available at fabric stores) work well. Secure the cheesecloth with pins; sew the nylon netting into a "pillowcase" shape and slip over each cooling rack.

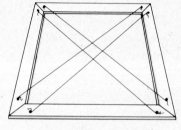

If you plan to dry a large quantity of food in the oven or the sun, you can construct wooden trays. They're easy to make and can be used for many years. You'll need lumber for the frame, kitchen string, and cheesecloth or stainless steel mesh. Do not use aluminum, fiberglass, copper, or galvanized screen; all of these materials will affect the quality of the food.

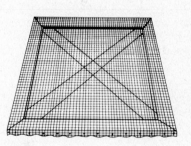

You can construct trays of any size, but keep these points in mind. For oven-drying, measure the interior of your oven to determine the dimensions of your trays; subtract at least an inch from width and depth measurements for air circulation. Trays for sun-drying can be as large as you can handle comfortably; the larger the tray, the more drying space it offers. For indoor drying, smaller stacking trays like those shown on page 67 are more convenient.

When you have built the frame, run strings diagonally across the frame, tacking them at opposite corners, to support the cheesecloth or stainless steel mesh. (See illustration.) Stretch and tack the cheesecloth or mesh to the other side of the frame.

SOLAR DRYERS

A solar dryer is a simple, home-built wooden frame topped with a glass panel tilted at an angle. The panel captures and intensifies the sun's heat, creating an "oven" effect. It shortens sun-drying time and helps protect food from insects, but it does have a few drawbacks. Because only a few trays will fit inside a solar dryer, the amount of food that can be dried at one time is limited. There is currently no effective way to control the temperature inside a solar dryer, making it difficult to control quality.

To dry fruits and vegetables in a solar dryer, follow the steps for sun-drying. If you'd like a plan for building a solar dryer, send a check for $2 to "Solar Dehydrator," American Plywood Association, Box 11700, Tacoma, WA 98411.

on stones, or on wood blocks—any sort of arrangement that raises the trays off the ground and allows air to circulate underneath them. To protect against insects, place 2×4s at opposite ends of the trays and tack cheesecloth to them; the cheesecloth should enclose the trays but should not touch the food. Turn the food several times each day so that it dries evenly. Bring the trays inside if you anticipate damp or foggy nights or morning dew.

After two days, start testing for doneness. (See pages 76 and 82–83.) Continue drying until food tests done.

In areas that have hot, dry breezes, you may dry food partially in the shade. In this method, called *sun-shade drying*, foods are dried in direct sun for the first two days and then transferred to the shade for the remainder of the drying time—several more days for fruits and vegetables. Food partially dried in the shade tends to dry more evenly and retain its color better than food dried entirely in the sun.

As a precaution against insect infestation during storage, fruits and vegetables that have been sun-dried should be *pasteurized* before storing. To pasteurize, seal dried foods in plastic food storage bags and place in the freezer at 0°F for at least two days. Or layer the fruit or vegetable loosely in roasting pans and bake in a preheated 175°F oven: 10 minutes for vegetables and 15 minutes for fruits. It's not necessary to pasteurize oven- or dehydrator-dried foods.

DRYING INDOORS

Today's homes generally do not offer the environmental conditions needed to dry most foods at room temperature. Exceptions are fresh herbs, flavorings, and flowers, which dry beautifully indoors in bunches or on trays. An enclosed porch, attic, or kitchen provides the ideal conditions—shade, good air circulation, and a warm temperature. Instructions for drying herbs, flavorings, and flowers indoors begin on page 84.

STORING DRIED FOODS

The final but critical steps in drying foods are testing for dryness and proper storage. As soon as dried food has cooled, test it for dryness following the guidelines on page 68, package it, and seal it tightly. Don't delay the packaging step, because the food will begin to reabsorb moisture at room temperature. Check packaged food for moisture during the first few weeks of storage. Any moisture inside the container is a sign that the food is not dry enough; it should be returned to the drying trays.

Store dried food in airtight, moisture-proof, and insectproof containers: glass jars (recycled mayonnaise, instant coffee, or pickle jars; canning jars), plastic food storage bags, plastic containers, cookie or cake tins, and coffee or shortening cans lined with plastic food storage bags.

Store dried foods individually in separate containers. Divide foods into small batches; the smaller the batch, the smaller the amount of food exposed to moisture each time the container is opened. And should a batch spoil, you will have lost a small amount, rather than a large quantity.

Label each container with the date the food was packed and the contents. Store in a cool, dry, dark place. The ideal storage temperature is 60°F or lower. Warmer temperatures cause dried food to lose flavor and nutrients. Darkness preserves fruit and vegetable color longer. A cupboard, pantry, or closet shelf or a covered cardboard box is ideal.

Stored at 60°F or lower, dried fruits and vegetables will keep up to one year. As the year progresses, color, nutrient value, and flavor will gradually diminish, so try to use dried food items with the earliest dates before more recently dried goods. Stored in the freezer, dried food will keep for years.

DRYING MEAT AND FISH FOR JERKY

The pioneers relied on jerky to sustain them during their arduous travels, but you can enjoy this treat today for its chewiness and concentrated, savory flavor. The recipes that follow simplify the commercial process by which jerky is made and adapt it for home-drying. Jerky makes a convenient, nutritious snack.

BEEF JERKY

One-half cup bottled teriyaki sauce may be substituted as the marinade in this recipe.

- 1 1/2 pounds beef round, chuck, or flank steak
- 1/4 cup soy sauce
- 2 tablespoons finely chopped onion
- 2 cloves garlic, minced
- 1 1/2 tablespoons brown sugar
- 1 tablespoon Worcestershire sauce
- 1 tablespoon grated fresh ginger root or 1/4 teaspoon ground ginger
- 1/8 teaspoon ground nutmeg

1. Cut meat across the grain in 1/8- to 1/4-inch-thick strips. (For easier slicing, partially freeze meat).

2. Combine meat strips and remaining ingredients in a plastic food storage bag or shallow dish. Close bag or cover dish with plastic wrap.

3. Refrigerate 6 hours or overnight to marinate meat.

4. *To dry in a dehydrator:* Drain marinade. Arrange meat in a single layer on trays. Dry at 120°F for 8 to 10 hours.

To dry in the oven: Drain marinade. Arrange meat in a single layer on trays. Dry at 120° to 140°F for 10 to 12 hours.

Sun-drying is not recommended.

5. When dry, jerky should feel firm, dry, and tough, but it should not crumble. There should be no moist spots. Store in airtight containers in a cool place, or refrigerate. Eat within 3 months. Jerky will keep 9 months if frozen.

Yield: About 6 ounces

SALMON JERKY

- 1 1/2 pounds skinless, boneless salmon fillets
- 1/4 cup soy sauce
- 2 tablespoons water
- 2 tablespoons brown sugar
- 1/8 to 1/4 teaspoon dried dillweed

1. Slice fillets across the grain in 1/8- to 1/4-inch-thick strips. (For easier slicing, partially freeze fish.)

2. Combine fish strips and remaining ingredients in a plastic food storage bag or shallow dish. Close bag or cover dish with plastic wrap.

3. Refrigerate 6 hours or overnight to marinate fish.

4. *To dry in a dehydrator:* Drain marinade. Arrange fish in a single layer on trays. Dry at 120°F for 12 to 15 hours. Blot with paper towels 3 or 4 times during drying to remove oil from fish.

To dry in the oven: Drain marinade. Arrange fish in a single layer on trays. Dry at 120° to 140°F for 10 to 12 hours. Blot with paper towels 3 or 4 times during drying to remove oil from fish.

5. When dry, jerky should feel firm, dry, and tough, but it should not crumble. Store in airtight containers or plastic food storage bags. Due to the high oil content of fish jerky, it has a shorter storage life than beef jerky: up to 2 weeks in the refrigerator or up to 3 months in the freezer.

Yield: About 6 ounces

Salmon fillets are thinly sliced, marinated, and dried to make a tasty and nutritious jerky.

LUFFAS

Luffas are grown and dried commercially to make sponges for bathing, but you can grow and dry them easily at home. Also known as vegetable sponges, dish-rag gourds, and Chinese okra, luffas grow to 10 to 15 feet at maturity, making cylindrical fruits 1 to 2 feet in diameter. To make sponges, pick the luffas at maturity and soak them in water for several days until the skin falls off. Then dry them in the sun.

DRYING FRUITS

Dried fruit makes a delicious, chewy snack or a naturally sweet addition to baked goods, salads, entrées, and vegetable dishes. Pack a handful of dried apple slices, pineapple wedges, or apricot halves in your knapsack for hiking or in the lunchbox for a midday treat. Preserve fresh summer blueberries to make a delicious bread when they're not available in the produce section, or add them to your own favorite recipes. Wake up to a bowl of crunchy granola with a sense of pride, knowing that you've made it yourself, or start the day with a compote of dried fruit. And try fruit leather treats for out-of-the-ordinary appetizers and snacks. You'll find recipes for all of these on pages 77–79.

Except for citrus fruits, melons, and some berries, all fruits can be dried successfully. Citrus fruits and melons, which have an extremely high water content when fresh, tend to be bland after their moisture has been evaporated by the drying process. You'll have better results drying only the citrus peel and enjoying melons fresh. Blackberries, boysenberries, and raspberries are not recommended for drying because their high seed content makes them unpalatable in dried form.

As discussed earlier, there are three basic methods for drying foods: in a dehydrator, in an oven, and in the sun. The drying method to use for fruits depends on climatic conditions where you live, the fruits you want to dry, and the kind of equipment you have on hand.

Each of the drying methods affects the appearance of fruit in a different way. Fruit dried in a dehydrator or an oven will be bright in color. If not sulfured first (see page 75), fruit dried in the sun will wrinkle and darken more than fruit dried by the other two methods. The degree of ripeness also affects the quality of the dried fruit. Use fully ripe, ready-to-eat fruit without blemishes or bruises. Underripe fruit will be less sweet than ripe fruit when dried. Overripe fruit takes longer to dry and may darken more than fruit at the peak of ripeness, and it will have a slightly fermented taste. See page 18 for information on where to obtain fresh fruits.

PREPARING FRUITS FOR DRYING

Gently wash fruit to remove any dirt. Prepare only as much fruit as your dehydrator, oven, or trays can accommodate at one time.

For fruit to dry quickly enough to avoid spoilage, it must be cut so that the moisture inside can escape. The simplest approach is to halve and pit most fruits. Although it's more time-consuming initially, slicing fruit speeds drying even more because it exposes more of the fruit's surface to the warm air. Small fruits that contain a pit, such as cherries, will dry faster if the pit is removed. You can peel fruit before halving or slicing it if you wish. To peel easily, dip whole fruit in boiling water for about 30 seconds, plunge into cold water, and slip off skins.

PRETREATING FRUITS BEFORE DRYING

Apples, apricots, bananas, peaches, nectarines, and pears darken when they are cut and exposed to air. To lessen this problem, pretreat these fruits before drying by any of the methods.

Which pretreatment technique to use depends on the drying method you choose. There are two basic ways to pretreat fruit: by dipping it in an antioxidant or a honey-water solution or by sulfuring it. You can use the first approach with any of the drying methods, although it will not maintain the quality of sun-dried fruits as well as sulfuring. Sulfur *only* fruits that will be sun-dried. (If you sulfur fruits, it's not necessary also to pretreat them with an antioxidant.) See page 74 for sulfuring instructions and safety precautions.

Antioxidant dips
Ascorbic acid dip. Dissolve 2 tablespoons ascorbic acid powder or crystals (available at drugstores) in 1 quart water, or use a commercial antioxidant (see page 22) following the instructions on the package. Dip cut fruit in solution; remove with a slotted spoon and drain well before loading onto drying trays.
Fruit juice dip. Dip fruit in undiluted pineapple juice or in a solution of ¼ cup

Most fruits lend themselves to drying. Pictured here are dried apples, pears, and persimmons.

SULFURING FRUITS

Always sulfur outdoors away from plants, shrubs, and trees. Sulfur dioxide fumes should not be inhaled. They can cause damage to both human and animal respiratory systems. Sulfuring must be done outdoors and sulfur-treated fruit must be dried in the sun—never in a dehydrator or an oven.

Recently, it has been found that some asthma sufferers are sensitive to sulfuring agents, including sulfur dioxide. If you give home-dried fruits to someone with this condition, alert them that your fruits have been treated with sulfur dioxide.

MATERIALS YOU'LL NEED

SULFUR. Use elemental sulfur—also called sublimed sulfur, Sulfur Flowers (USP standard), or flowers of sulfur. It is free of impurities, burns readily, and may be purchased at most pharmacies. Sulfur preparations for garden dusting are more difficult to ignite than elemental sulfur.

CLEAN METAL CONTAINER. To hold the sulfur. A 1-pound coffee can or an aluminum pie tin will be adequate for small amounts of fruit.

PLASTIC OR WOODEN TRAYS. See page 69 for a description of appropriate trays.

WOODEN BLOCKS. To separate stacked trays.

FIRE BRICKS OR CONCRETE BLOCKS. To raise the stack of trays 6 inches higher than the container of burning sulfur.

BOX. Of heavy cardboard or of wood. It must be large enough to fit over the stacked trays with 1½ inches of clearance at the top and on three sides. You'll need 12 inches of clearance at the front or on one side for the sulfur container. The bottom tray should be 6 inches above the burning sulfur. A large carton of the type in which household appliances are shipped works well. Seal any cracks or openings with tape.

HOW TO SULFUR

1. To determine how much sulfur to burn, first halve or slice fruit and weigh it. The amount of sulfur to use depends on the length of time the fruit is to be sulfured (see chart on page 76 for specific fruits), the weight of the fruit, and the dimensions of the box you are using. Generally, if you use a cardboard box to cover the trays, allow 1 tablespoon of sulfur per pound of prepared fruit. If you construct a sulfuring box from wood, you'll need only half that amount per pound of fruit because the box will be more airtight.

2. Spread fruit on trays in a single layer, pit cavity or cut surface up. Pieces should not touch each other. Sulfur only one kind of fruit at a time because different fruits require different sulfuring periods.

3. Select a gravel or dirt area. Position fire bricks to support each end of the first tray so that it will be at least 6 inches above the metal container that will hold the sulfur. Continue stacking trays, placing wooden blocks at their corners to separate them by at least 1½ inches.

4. Remove the top of the box and turn the box upside down. Cut a 6-inch door near one of the bottom corners of the box; this allows air to enter so that the sulfur will burn. On the side of the box opposite the intake door, cut a 6-inch slash near the top.

5. Measure the sulfur and place it in the metal container in a layer not more than ¼ inch deep. Place the container at the side or in front of the stack of trays. Do not place it under the stack; this exposes the bottom tray to too much sulfur dioxide.

6. Light the sulfur and position the container. (Do not leave used matches in the container.) Place the box, open end down, over the trays and immediately seal the bottom edges of the box by pushing dirt against them. Begin timing.

7. When sulfur is brownish in color, has melted to a thick, sticky consistency, and is burning well with a clear blue flame (10 to 15 minutes), tightly seal both the door and the slash in the box. When the time is up, remove the box and check the fruit. It is properly sulfured when it is bright and glistening and a small amount of juice appears in the pit cavity. If the fruit is not finished, replace the box and continue the process. Use care when removing the sulfur box so that fumes do not blow in your direction. (A protective mask of the sort painters and carpenters use is a sensible precaution.)

8. Transfer trays to the sun for drying.

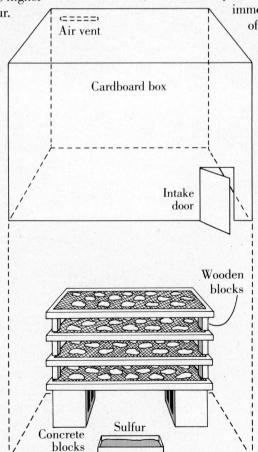

Air vent

Cardboard box

Intake door

Wooden blocks

Concrete blocks

Sulfur

lemon juice to 1 quart water. Drain well before drying.

Honey-water dip. Dissolve 1 cup sugar in 3 cups hot water. Cool to lukewarm and add 1 cup honey; stir well. Dip fruit in honey-water solution; drain well before arranging on trays. Rhubarb, pineapple, bananas, and apricots take on a sweet, almost candylike flavor when dipped in a honey-water solution before drying.

Sulfuring. Before fruits are dried commercially, they commonly undergo a process called sulfuring to inhibit mold, retain natural, bright color, help protect nutrients, and lessen the chance of insect infestation during drying. In this process, the fruit is exposed to sulfur dioxide fumes in a closed chamber. Sulfuring actually bleaches the fruit.

All unsulfured fruits darken when exposed to the sun. Raisins are a case in point. Golden raisins are made from Thompson seedless grapes, picked from the vine, treated with sulfur dioxide, and dried indoors. Dark raisins are also made from Thompson seedless grapes, but are sun-dried without any sulfur treatment.

The color and quality of sun-dried fruit will be better if it is sulfured before drying, but sulfuring is not mandatory. You can successfully dry fruit in the sun without sulfuring, but you should pretreat fruit that darkens with an antioxidant dip.

See the opposite page for sulfuring instructions and safety precautions.

How to Dry Fruit

Arrange prepared fruit on drying trays skin side down, leaving just enough space between pieces so that they do not touch. Drying pieces of similar size together saves sorting time toward the end of the drying period. Check the chart on page 76 for suggested drying methods, times, and signs of dryness for specific fruits.

Keep an eye on fruit toward the end of the drying time and test frequently to avoid overdrying. When fruit is completely dry, cool; package and store following the guidelines on page 70.

QUICK AND EASY DRIED FRUIT TREATS

PEANUT BUTTER–BANANA SNACKS. Sandwich a dab of peanut butter between two banana chips.

FRUIT-FLAVORED YOGURT. Purée in blender until smooth 1⅓ cups dried blueberries, cherries, apples, peaches, or pears and 1 cup water. Stir purée into 2 cups plain yogurt. Sweeten with brown sugar or honey, and add cinnamon, nutmeg, or other spices to taste.

APPLESAUCE. Purée in blender until smooth 1½ cups dried apple slices, 1½ cups water, and 1 to 2 teaspoons lemon juice. Sweeten with honey or sugar, and add cinnamon, nutmeg, or other spices to taste.

FRUIT SMOOTHIE. In a blender, whirl together choice of dried fruit, milk or plain yogurt, unsweetened apple or pineapple juice, crushed ice, and sugar or honey to taste.

BERRY SAUCE. Purée 1 cup berries with warm water in a blender or food processor. Use as a topping for ice cream or to flavor shakes and smoothies.

SPIRITED FRUIT. Soak dried grapes (raisins) and other dried fruit in rum or brandy until plump. Spoon spirited fruit over ice cream or use in baking.

COOKING WITH DRIED FRUIT

If a recipe calls for chopped dried fruit, it will be easier to chop if you toss 1 cup fruit with 1 teaspoon oil and chop with a French knife; or freeze fruit and then chop. When a recipe calls for plumped dried fruit, cover the amount needed with very hot tap water and soak 2 to 5 minutes. (Longer soaking results in flavor and nutrient loss.) For variety in your baked goods, sauces, toppings, and desserts, soak dried fruit in a favorite fruit juice or liqueur. Cover the fruit with the liquid; soak overnight. Delicious recipes using home-dried fruit begin on page 77.

GUIDE FOR DRYING FRUITS

The times in this table should be used as guidelines only. Drying time depends to some extent on the dehydrator or oven setting or the outdoor temperature, the amount of moisture in the food, and the humidity in the air. Watch fruits carefully toward the end of drying time and test frequently for correct texture and dryness.

FRUIT	PREPARATION	APPROXIMATE SULFURING TIME BEFORE SUN-DRYING	APPROXIMATE DRYING TIME		INDICATORS OF DRYNESS
			SUN	DEHYDRATOR/OVEN (120°–140°F)	
Apples*	Wash and core; peel if desired. Cut in ¼-inch slices or rings.	45–60 minutes	3–4 days	6–12 hours	Soft and pliable; no moisture in center when cut.
Apricots*	Peel if desired. Halve or slice, removing pit.	2 hours (halves); 1 hour (slices)	2–3 days	16–36 hours (halves); 7–10 hours (slices)	Same as for apples.
Bananas*	Peel and cut in ¼-inch slices.	—	2–3 days	8–16 hours	Leathery but still chewy. (Longer drying will turn banana slices into crisp chips.)
Blueberries; Cranberries	Halve.	—	2–4 days	8–12 hours	Leathery but still chewy.
Cherries	Pit and halve.	—	1–2 days	18–30 hours	Leathery but still chewy.
Figs	Peel and quarter.	—	4–5 days	10–12 hours	Pliable; slightly sticky but not wet.
Grapes	Halve; seed if desired.	—	3–5 days	24–48 hours	Raisinlike texture; pliable; chewy.
Peaches;* Nectarines*	Peel if desired. Halve or cut in ¼-inch slices, removing pits.	2–3 hours (halves or slices)	3–5 days (halves or slices)	24–36 hours (halves); 8–12 hours (slices)	Same as for apples.
Pears*	Halve and core, or core and cut into ¼-inch slices.	5 hours (halves or slices)	5 days (halves or slices)	24–36 hours (halves); 10–14 hours (slices)	Same as for apples.
Persimmons	For Fuyu variety, select firm fruit; for Hachiya variety, let fruit ripen until soft. Peel and cut into ¼-inch slices.	—	5–6 days	14–18 hours	Light to medium brown; tender but not sticky.
Pineapple	Peel, core, and cut crosswise into ¼-inch slices. Dry slices whole or cut them in wedges.	—	4–5 days (slices) 3–4 days (wedges)	24–36 hours (slices) 18–24 hours (wedges)	Chewy and dry to center.
Plums	Halve or cut in ¼-inch slices, removing pit.	—	4–5 days	18–24 hours (halves); 8–10 hours (slices)	Fairly hard and leathery but still chewy.
Rhubarb	Cut in ¼-inch slices.	—	2–3 days	18–20 hours	Hard to crisp.
Strawberries	Halve or cut in ¼-inch slices.	—	1–2 days	20 hours (halves); 12–16 hours (slices)	Leathery but still pliable.

*Pretreat these fruits to protect their color. Dip in an antioxidant or a honey-water solution (see page 72) if dehydrator- or oven-drying. For sun-drying, sulfur fruits (see page 74) or dip in solution.

DRIED FRUIT COMPOTE

This is a typical Eastern European dessert that is also delicious for breakfast. Add sugar to taste.

1 pound mixed dried fruits

2 cups water

1/4 to 3/4 cup sugar

1 cinnamon stick

6 cloves

Juice and grated peel of 1 lemon

1. Place dried fruits in a large bowl. Add water and let soak several hours.
2. Drain soaking liquid into a medium saucepan. Add sugar, cinnamon stick, and cloves and heat, stirring, to dissolve sugar. Remove from heat; add lemon juice and peel.
3. Pour syrup over fruit. Cool to room temperature.
4. Cover and chill before serving.

Yield: About 6 servings

BLUEBERRY BREAD

Reconstitute blueberries, dried in the summer, to make this delicious bread.

1 cup dried blueberries

4 tablespoons butter

3/4 cup sugar

1 egg

2 cups all-purpose flour

2 teaspoons baking powder

1/2 teaspoon salt

1/2 cup milk

1. Place dried blueberries in a small bowl. Cover with warm water and let stand about 30 minutes to reconstitute. Drain.
2. In a mixing bowl, cream butter and sugar until light and fluffy. Beat in egg.
3. Sift together flour, baking powder, and salt.

4. Add flour mixture to creamed mixture alternately with milk, mixing well after each addition. (Batter will be thick.) Fold in blueberries.
5. Pour batter into a greased 9- by 5-inch bread pan. Bake at 350°F for about 1 hour and 10 minutes, until pick inserted in center comes out clean. Cool bread in pan.

Yield: 1 loaf

OLD-FASHIONED GRANOLA

1/3 cup honey

1/4 cup brown sugar

3 cups oatmeal

1 cup wheat germ

1/2 cup bran

1/2 cup whole wheat flour

1/2 cup soy flour

1 cup shredded coconut

1/2 cup unsalted sunflower seeds

1/2 cup sesame seeds

1/3 cup vegetable oil

3/4 cup dried grapes or blueberries

1/4 cup chopped dried apple

1/4 cup chopped dried pineapple

1/4 cup chopped dried cherries

1/4 cup chopped dried bananas

1. Preheat oven to 325°F. In a large bowl, mix together honey, brown sugar, oatmeal, wheat germ, bran, whole wheat flour, soy flour, coconut, sunflower seeds, sesame seeds, and vegetable oil.
2. Spread not more than 1 inch deep on 1 or 2 shallow baking sheets and bake for 1 hour, stirring often.

Snack mixes make portable, packable treats for hikers and backpackers.

3. Stir in dried fruits.
4. Cool and store in airtight containers. For long-term storage, keep in refrigerator.

Yield: 7 cups

CAROB-NUT SNACK MIX

1/2 cup whole unblanched almonds

1/2 cup cashews

1/2 cup shredded coconut

1/2 cup dried grapes, cherries, or blueberries

1/2 cup carob chips

1/4 cup sunflower seeds

1/4 cup chopped dried apricots or peaches

Mix all ingredients together.

Yield: 3 cups

FRUIT-NUT SNACK MIX

3/4 cup dried grapes, cherries, or blueberries

1/2 cup chopped dried pears

1/2 cup chopped dried apples

1/2 cup cashews

1/2 cup shredded coconut

1/2 cup whole unblanched almonds

1/4 cup chopped dried pineapple

Mix all ingredients together.

Yield: 3 1/2 cups

PEANUT-CHOCOLATE SNACK MIX

2 cups shelled peanuts

1 cup bite-size candy-coated chocolate bits (such as plain M&M candies)

1 cup bite-size candy-coated peanut butter bits (such as Reese's Pieces)

1 cup whole unblanched almonds

1/2 cup chopped dried banana

Mix all ingredients together.

Yield: 5 1/2 cups

PEACH CREAM-CHEESE PINWHEELS

Children will have fun making these tasty confections.

1 Peach Leather (recipe on opposite page)
1 package (8 oz) cream cheese, softened
1 tablespoon honey
1 tablespoon milk

1. Prepare Peach Leather as directed on opposite page.
2. Beat cream cheese with honey and milk. Spread evenly over leather.
3. Roll up jelly roll fashion from the short end. Slice into ½-inch pieces.

Yield: 14 to 16 pinwheels

FRUIT LEATHERS

If you've ever tried the colorful commercial "fruit rolls," you know how deliciously chewy these translucent fruit snacks are. Making fruit rolls, commonly known as fruit leathers, at home is a fun, economical, and easy project: Simply purée fruits and dry slowly in the sun, an oven, or a dehydrator. They'll turn into glistening strips of fruit "candy."

SETTING UP

To make fruit leathers, you'll need drying trays or baking sheets, plastic wrap, and a blender or food processor. Line trays or baking sheets with plastic wrap, taping it to the underside of the trays. The fruit purée tends to stick to waxed paper or aluminum foil. A light coating of nonstick vegetable spray will help prevent this.

MAKING THE PURÉE

You can make fruit leathers from fresh, frozen, or well-drained canned fruit. Fresh or frozen fruit should be ripe or slightly overripe. In fact, leathers made from slightly overripe fruit will have a more intense flavor. Choose apricots, apples, peaches, nectarines, plums, strawberries, cherries, pears, pineapples, raspberries, blueberries, cranberries, or tomatoes. Citrus fruits are not suitable because they do not contain enough pulp.

To begin, wash fruit and pat dry. Peel and core if necessary. Fruits like peaches and nectarines do not have to be peeled, but leathers made without skins are lighter in color and have a smoother texture.

Purée prepared fruit in a blender or food processor until smooth, adding 1½ teaspoons lemon juice to each quart fruit purée. To make one sheet of leather measuring approximately 9 inches by 12 inches, you'll need about 1½ cups fruit purée (4 to 5 medium peaches). If you're puréeing a fruit like apples, you may need to add a little water or fruit juice (apple, pineapple) to get the blending process started, but add only enough water to facilitate blending. The purée should be thin enough to pour but not too runny. If it's too thin, cook it slowly over low heat to evaporate excess moisture.

Taste the purée for sweetness, and add honey or sugar to taste if it's too tart. Sweetness intensifies as the leather dries, so add honey or sugar sparingly.

FORMING THE SHEET

Pour the purée onto the plastic-lined drying trays or baking sheets. Using a rubber spatula, spread purée *evenly* to a thickness of ⅛ to ¼ inch. If the purée is thin at the edges and thick at the center, the leather will be brittle around the edges and sticky in the middle. The thicker the purée is spread, the longer it takes to dry.

DRYING THE LEATHER

Sun-drying. Refer to the information on page 69 on sun-drying. Place filled trays in full sun. Protect purée from insects by suspending cheesecloth over, but not on, the surface of the trays, as directed on page 70. Drying time depends on the temperature, humidity, and wind; it will vary from 8 hours to 3 days. Test often and don't leave in the sun longer than necessary.

Oven-drying. Refer to the information on page 68 on oven-drying. Preheat an electric oven to its lowest setting (no higher than 140°F). Leave oven door ajar 3 to 4 inches and dry fruit purée 6 to 8 hours. In a gas oven, the pilot light will supply sufficient warmth. Open the door every few hours to let moisture escape. Leathers will dry in 1 to 2 days in a gas oven.

Dehydrator-drying. Refer to the information on page 66 on dehydrator-drying. Dehydrate at 120°F for 8 to 12 hours.

TESTING FOR DONENESS

Properly dried fruit leather is pliable and "leatherlike," firm to the touch, and no longer sticky. Leathers are dry when they can be peeled easily from the plastic wrap. If underside feels slightly sticky, invert leather (and plastic) onto tray or baking sheet so that sticky side faces up, remove plastic, and continue drying. Times given above are estimates; amount of moisture in the purée and depth of the purée affect drying time.

WRAPPING AND STORING LEATHERS

To wrap fruit leathers, simply roll up sheets with plastic wrap attached, tucking in edges of wrap as you roll. Overwrap each roll in a second sheet of plastic wrap for added protection against moisture. Package in plastic bags or airtight containers. Store up to 6 months at room temperature in a cool, dry, dark place or up to a year in the refrigerator or freezer.

A Few Fruit Leathers

Here are some ideas to try, and a few recipes for specific leathers.

PEACH LEATHER. Purée until smooth 1½ cups peeled, chopped peaches, 2 teaspoons lemon juice, 1 tablespoon (or more to taste) honey, and ⅛ teaspoon ground coriander. Makes 1 sheet leather.

STRAWBERRY LEATHER. Prepare as for peach leather, substituting 1½ cups chopped strawberries for the peaches and ⅛ teaspoon ground mace for the coriander.

PLUM LEATHER. Prepare as for peach leather, substituting 1½ cups peeled, chopped plums for the peaches and ⅛ to ¼ teaspoon ground cinnamon for the coriander.

SPICY LEATHER. To each 1½ cups purée, add ⅛ to ¼ teaspoon of a favorite spice (for example, coriander, cinnamon, allspice, cloves, nutmeg, or ginger). The spice flavor will concentrate during drying, so use less rather than more.

NUTTY LEATHER. When leather is partially dry but still moist, sprinkle chopped nuts over top.

TWO-TONED LEATHER. Prepare two separate batches of fruit purée using two different fruits. Delicious combinations include apple-pear, peach-plum, raspberry-cherry, and apricot-pineapple. Pour equal amounts of the two purées on each plastic-lined tray or baking sheet. Using a rubber spatula, gently swirl them together for a marbled effect.

DOUBLE-FRUIT LEATHER. Prepare two separate batches of fruit purée using two different fruits. Combine in blender or food processor before pouring onto prepared trays or baking sheets.

FILLED LEATHER. To make delicious appetizers, snacks, or healthy candies, simply spread dried leather with a favorite filling—softened cream cheese, peanut butter, and commercial cheese spreads are just a few of the possibilities. Roll up jelly roll fashion and slice into bite-size pieces. Start with the recipe for Peach Cream-Cheese Pinwheels on the opposite page.

FRUIT SAUCE FROM FRUIT LEATHER. Purée 1 cup broken pieces of fruit leather and 1 cup water in blender until smooth. Serve as a sauce for ice cream, cake, pancakes, or waffles.

Fruit leathers roll up to make healthful fruity snacks, perfect for lunch boxes and after-school munching.

Cooking with Dried Vegetables

The water removed from vegetables during drying must be replaced by soaking, cooking, or a combination of the two. Depending on the kind and the thickness, dried vegetables can take anywhere from 20 minutes to 2 hours to rehydrate. Begin by soaking dried vegetables in a small amount of boiling water or broth for 20 to 30 minutes, stirring occasionally, until the vegetables no longer look shriveled and have absorbed most of the liquid. If after 30 minutes the vegetables are still shriveled but have absorbed the liquid, add more boiling liquid. When most of the shriveled look is gone, use the vegetables in recipes as you would fresh.

Drying Vegetables

We're most familiar with dried vegetables for their roles in commercially packaged soups and sauces, flavoring packets for convenience foods, and lightweight meals for camping and backpacking. Considering the variety of fresh and frozen vegetables available in supermarkets today and the fact that the taste and texture of many rehydrated dried vegetables does not equal that of the frozen or canned version, you may decide to dry vegetables only to create your own convenience foods; to use as ingredients in soups, stews, casseroles, stuffings, and sauces; and as an economical alternative to purchased camping foods.

Some vegetables—notably asparagus, broccoli, and cauliflower—do not rehydrate well; they'll taste far better if preserved by freezing. Others—carrots and onions, for example—are available the year around at reasonable prices, so you may choose not to dry them. And keep in mind that most dried vegetables—corn is an exception—are not at their best when cooked and served plain as a side dish. For some ideas on using dried vegetables in cooking, see the recipes on the opposite page.

For the best quality product, dry only tender vegetables. Overmature vegetables will be tough and woody and immature vegetables will have poor color and flavor when dried. Remember, for the best flavor pick early and dry very shortly after picking. (See page 18 for information on where to obtain fresh vegetables.) Store vegetables in the refrigerator until you are ready to dry them.

Vegetables are dried until only about 5 percent of their moisture remains. Although they can be dried by any of the three methods—in a dehydrator, an oven, or the sun—in general, dehydrator- and oven-drying are preferred to sun-drying because vegetables are susceptible to spoilage during the long sun-drying process.

Preparing Vegetables for Drying

Wash vegetables carefully in cold water and pat dry. Most will need to be peeled, trimmed, cored, and sliced or shredded. Cut vegetables in pieces of uniform size so that they will dry at the same rate. Drying time depends on the thickness of the pieces; the longer the drying period, the less tender the dried vegetable, so it's best to slice vegetables thinly. Prepare only as many vegetables as you can dry at one time.

Pretreating Vegetables Before Drying

Most vegetables need to be pretreated by blanching before drying to inactivate the enzymes that cause ripening and eventual decay. (See the chart on pages 82 and 83 for those that can be dried without blanching.) Blanching means heating over steam or in boiling water. Steam-blanching preserves more of the vitamin and mineral values of the food, but it takes longer than water-blanching. It also requires that you stir the food occasionally to ensure that the steam reaches all of the pieces. Instructions for both blanching methods are on page 28, and the chart on pages 82 and 83 lists blanching times.

How to Dry Vegetables

Arrange prepared vegetables on drying trays, leaving space between pieces. Dry pieces of similar size on the same tray. With the exception of onion, garlic, and peppers, which tend to impart their flavors to milder vegetables, you can dry various kinds of vegetables together in a dehydrator or an oven.

Controlling temperature and drying time is critical to the quality of dried vegetables. If the temperature is too low, vegetables may spoil before they dry completely. If the temperature is too high, vegetables will have a "cooked" taste and be less nutritious and tender.

Test for dryness toward the end of the drying period. Most vegetables are brittle, crisp, or hard when thoroughly dry. Specific dryness indicators are given in the chart on pages 82 and 83.

Cool dried vegetables before packing for storage. During the first few weeks of storage, check containers for moisture once or twice a week. One small piece that is still moist can cause the entire batch to mold. This alone is good reason to store dried food in small batches.

CREAMED CORN

This vegetable side dish is a Midwestern favorite.

- 1 1/2 cups dried corn kernels
- 3 cups boiling water
- 3/4 cup half-and-half (light cream)
- 1 teaspoon sugar
- 1/4 cup butter or margarine
  Salt and ground black pepper

1. Place dried corn in a medium saucepan and pour boiling water over it. Let stand 2 hours.
2. Add half-and-half, sugar, and butter.
3. Cover and barely simmer for 1 hour or longer, until corn is soft. Watch carefully toward the end of the cooking period so that corn does not boil dry and burn. Season with salt and pepper to taste.

Yield: 4 to 6 servings

VEGETABLE SOUP MIX

This dehydrated soup mix is great for camping.

- 2 teaspoons instant chicken-flavored bouillon
- 1/4 cup dried carrot slices
- 1/4 cup dried celery slices
- 1/4 cup dried diced green pepper
- 1/8 teaspoon dried thyme

1. Combine all ingredients in a small jar or plastic food storage bag.
2. To prepare soup, stir mixture into 2 cups boiling water. Reduce heat, cover pot, and simmer 15 to 20 minutes.

Yield: Two 1-cup servings

MUSHROOM-BARLEY SOUP MIX

- 1/2 cup dried sliced mushrooms
- 1/4 cup pearl barley
- 2 teaspoons instant chicken-flavored bouillon
- 1/4 teaspoon dried thyme

1. Combine all ingredients in a small jar or plastic food storage bag.
2. To prepare soup, stir mixture into 4 cups boiling water. Reduce heat, cover, and simmer 35 to 45 minutes.

Yield: Two 1-cup servings

PUMANTE

You can make your own *pumante* (Italian dried tomatoes in oil) for a fraction of what it costs in specialty food shops. Use the best quality olive oil you can afford. The dried tomatoes keep at room temperature for as long as the oil remains fresh. Refrigerate for longer storage. Add *pumante* to salads, serve as an appetizer, toss in pasta, or use as an attractive edible garnish.

- 3 pounds Roma or other small pear-shaped tomatoes
  Salt
- 2 or 3 cloves garlic, peeled
- 2 or 3 small sprigs rosemary
  Olive oil

1. Choose the smallest tomatoes you can find. Slice lengthwise almost in half and lay them open like a book (cut side up). Sprinkle cut surfaces lightly with salt.
2. Place tomatoes, cut side up, on dehydrator or other drying trays. Dry in a dehydrator or the oven at

120° to 140°F: 4 hours or longer in a dehydrator; 24 hours or longer in the oven.
3. When tomatoes are dry, they will have shriveled to small, flattish ovals and will feel dry but still pliable—not brittle—to the touch.
4. Pack tomatoes, garlic, and rosemary loosely into 2 or 3 half-pint jars. Pour in enough oil to cover tomatoes completely—they may mold if exposed to air. Cap jars.
5. Let stand in a cool, dark place for 1 month for flavors to develop.

Yield: 2 or 3 half-pints

Pumante—dried tomatoes in oil—is a flavorful addition to salads or pasta, and it's not difficult to make in a dehydrator or in your oven.

GUIDE FOR DRYING VEGETABLES

The times in this table should be used as guidelines only. Drying time depends to some extent on the efficiency of the heat source, the amount of moisture in the food, and the humidity in the air. Watch vegetables carefully toward the end of drying time and test frequently for correct texture and dryness. Times are based on a dehydrator or oven temperature of 120° to 140°F.

VEGETABLE	PREPARATION	APPROXIMATE BLANCHING TIME		APPROXIMATE DRYING TIME		INDICATORS OF DRYNESS
		METHOD	TIME (MIN.)	METHOD	TIME (HRS.)	
Asparagus*	Cut into ½-inch pieces.	Steam Water	4–5 3½–4½	Dehydrator or oven	8–10	Brittle
Beans, green	Cut in short pieces or lengthwise.	Steam Water	2–2½ 2	Dehydrator or oven	8–14	Brittle
Beets	Cook as usual; cool; peel. Cut in ⅛-inch slices.	No blanching required		Dehydrator or oven	8–12	Tough; leathery
Broccoli*	Trim; slice stalks lengthwise no more than ½-inch thick.	Steam Water	3–3½ 2	Dehydrator or oven	12–18	Brittle
Brussels sprouts	Cut in half lengthwise.	Steam Water	6–7 4½–5½	Dehydrator or oven	12–24	Hard to brittle
Cabbage	Remove outer leaves; quarter and core. Cut into ⅛-inch thick slices.	Steam Water	2½–3 1½–2	Dehydrator or oven	10–15	Tough to brittle
Carrots	Cut off roots and tops. Peel; cut in ⅛-inch slices.	Steam Water	3–3½ 3½	Dehydrator or oven	10–18	Tough to brittle
Cauliflower*	Break into small flowerets.	Steam Water	4–5 3–4	Dehydrator or oven	12–15	Tough to brittle
Celery	Trim stalks and cut in ¼-inch slices.	Steam Water	2 2	Dehydrator or oven	10–18	Crisp; brittle
Corn, cut	Cut kernels from cob after blanching.	Steam Water	2–2½ 1½	Dehydrator or oven Sun	6–12 2–3 days	Brittle; crunchy
Eggplant	Trim and cut in ¼-inch slices.	Steam Water	3½ 3	Dehydrator or oven	12–24	Brittle
Mushrooms	Remove and discard any tough, woody stems. Trim ⅛ inch off stems. Slice.	No blanching required		Dehydrator or oven	8–12	Tough; leathery

*These vegetables do not rehydrate well; choose freezing over drying if possible.

Vegetable	Preparation	Approximate Blanching Time		Approximate Drying Time		Indicators of Dryness
		Method	Time (min.)	Method	Time (hrs.)	
Okra	Trim and slice crosswise in ⅛- to ¼-inch slices.	No blanching required		Dehydrator or oven	8–10	Tough to brittle
Onions	Remove outer skin, top, and root end. Cut into ⅛- to ¼-inch slices.	No blanching required		Dehydrator Oven	10–20 Not recommended	Brittle and papery
Peas	Shell.	Steam Water	3 2	Dehydrator or oven Sun	12–16 2–3 days	Wrinkled and hard
Peppers, green or red	Stem and core. Cut crosswise into ¼-inch circles or lengthwise into ¼-inch strips.	No blanching required		Dehydrator or oven	8–12	Flexible; dry to the touch
Parsnips (See carrots.)						
Potatoes, white and sweet	Peel. Cut in ⅛-inch slices or ¼-inch strips.	Steam Water	6–8 5–6	Dehydrator or oven	8–12	Brittle
Spinach and other greens	Trim.	Steam Water	2–2½ 1½	Dehydrator or oven	8–14	Brittle
Squash, summer	Trim and cut into ¼-inch slices.	Steam Water	2½–3 1½	Dehydrator or oven	6–10	Brittle
Squash, winter	Peel and cut into 2- to 4-inch pieces, ¼-inch thick.	Steam Water	2½–3 1½	Dehydrator or oven	10–16	Crisp; hard
Tomatoes	Cut in ¼-inch-thick slices.	No blanching required		Dehydrator or oven	12–18	Crisp; brittle
Roma, or other small pear-shaped	Cut in half lengthwise.	No blanching required		Dehydrator or oven	15–24	Flexible; dry to the touch

Herb Blends

You may combine herbs to make herb blends before packaging. For starters, try one of the suggestions below. Each makes about ½ cup. Package blends in small glass jars or airtight plastic containers.

BOUQUET GARNI (for soups and stews). Mix together 3 tablespoons *each* dried thyme, dried parsley, and crumbled dried bay leaves.

ITALIAN HERB SEASONING (for spaghetti or pizza sauce). Mix together 3 tablespoons *each* dried oregano and dried basil and 2 tablespoons crumbled dried bay leaves.

HERB POULTRY SEASONING (for roasted or broiled poultry and for stuffings). Mix together 3 tablespoons dried rosemary and 2 tablespoons *each* dried sage and dried thyme.

DRYING HERBS

Fresh herbs are one of the easiest foods to dry: at room temperature; in a dehydrator; or in a conventional, microwave, or convection oven. The only method not recommended is sun-drying, because hot, direct sun diminishes the delicate flavor, color, and aroma of herbs.

No matter which drying method you choose, a very low temperature and good air circulation are important to ensure that herbs dry well and to preserve flavor. Drying concentrates the essential oils of herbs, which provide the flavor in cooking. The flavor of dried herbs is more pronounced than that of fresh. For the same flavor as 1 tablespoon fresh herbs, use 1 teaspoon dried.

Some herbs retain their natural flavors better than others during drying. Among those that dry well are bay, thyme, rosemary, marjoram, savory, sage, oregano, and tarragon. Chervil, parsley, and chives are best frozen.

Leafy herbs are easier to dry if the leaves are left on the stems; strip them off when thoroughly dry, and pack for storage. If you dry herbs in an oven or a dehydrator, dry types separately to eliminate flavor exchanges.

PREPARING HERBS FOR DRYING

Rinse herbs in cool water; remove any discolored or dead leaves or stems. Shake off excess water and dry on paper toweling. Then choose the drying method that's most convenient for you. Herbs are dry when they crumble easily and their stems snap when bent; seeds should be brittle. *Note:* Although seeds will drop from their pods during the drying process, they are not completely dry until they are brittle.

HOW TO DRY HERBS

To dry at room temperature. Hang herbs by their stems in bunches from the ceiling (called bunch-drying) or lay herbs flat on trays. Dry away from direct sunlight: An attic, covered porch, or kitchen that stays at 65° to 90°F is a good location. If you dry herbs in a covered area outdoors where dew collects, bring them indoors overnight.

Bunch-drying is ideally suited for herbs with long stems, like marjoram, rosemary, and sage. Tie bunches at the stem ends; hang upside down to dry. Bunches may be dried in small brown paper bags to keep dust from collecting on them and to catch seeds. Gather bag opening around stems and tie; herbs should hang freely inside bag. Cut several ½-inch holes in bags and suspend from the ceiling at varying heights to increase air circulation. Leaves and seeds will be thoroughly dry in a week or two.

Tray-drying works well for herbs with large leaves, like basil, and herbs with stems too short to hang. Use trays of any size. Spread leaves or stems in a single layer. To protect against insects and dust, cover with cheesecloth. Turn leaves or stems each day or two. When leaves crumble easily (about a week), herbs are dry.

To dry in a dehydrator. Spread herbs or seeds in a single layer on dehydrator trays and dry at 90° to 100°F for 2½ to 10 hours. Do not dry herbs with foods that must be dried at a higher temperature. Dehydrator-drying is virtually guaranteed to preserve the flavor and color of herbs.

To dry in a microwave oven. Place four or five stems of herbs on a double thickness of paper towels; cover with a single layer of toweling. Microwave at full power (high) until leaves are brittle: about 2 minutes for small leaves; 3 minutes for large. If leaves are not yet brittle, microwave an additional 30 seconds.

To dry in an oven or a convection oven. Use this method only if the oven can maintain a temperature of 120°F or lower; electric and convection ovens may not be able to maintain this low a temperature. The pilot light in a gas oven often provides enough warmth to dry herbs. Arrange herbs in a single layer on trays. Keep oven door ajar about ½ inch to allow moisture to escape. Watch the herbs carefully because they will dry quickly (1 to 3 hours).

STORING DRIED HERBS

When herbs are crumbly and feel dry, remove the leaves from the stems. Whole leaves keep their flavor longer than those that are crumbled before storage. Package in small glass jars. Inspect during the first few weeks for moisture. If condensation appears, redry. Label and date containers. Store in a cool, dry, dark place. Properly stored, dried herbs retain their flavor and color for up to a year.

It's simple to dry fresh garden herbs indoors—just hang them in bunches by their stems.

ORANGE-FLAVORED COFFEE

2 oranges

1/2 cup powdered instant coffee (not crystals)

3/4 cup instant cocoa mix

3/4 cup powdered instant nondairy creamer

1. With vegetable parer, peel oranges in continuous spirals. Place peel on baking sheet and dry in 120° to 140°F oven for 2 to 6 hours. Or dry in a dehydrator at 120°F for 2 to 6 hours.

2. Place dried peel and remaining ingredients in a jar; cap. Store mix 1 week for flavors to blend.

3. To serve, place 2 level tablespoons mix in each cup. Add ¾ cup boiling water; stir well.

Yield: 2 cups mix (about 16 servings)

VARIATIONS

Cinnamon-orange coffee: Add 2 teaspoons ground cinnamon to dry mix.

Clove-orange coffee: Add ¼ teaspoon ground cloves to dry mix.

DRYING FLAVORINGS

Common flavorings like citrus peel, chiles, garlic, and shallots are easy to dry. Use them as you would their fresh counterparts; but because they are storehouses of concentrated flavors, you'll need less to provide the same flavor as the fresh form.

HOW TO DRY CITRUS PEEL

Drying orange, lemon, or lime peel is a fun, easy project that even children can try. Depending on the drying method you choose, you can dry it in strips or grate it before drying. Dried peel adds refreshing flavor to cakes, pies, breads, cookies, fruit salads, and whipped cream. It can also be used to flavor coffee or pepper. (See left and page 88 for recipes.) Dry peels at room temperature, in a dehydrator, or in a conventional or microwave oven.

Preparation. Wash fruit and pat dry. Using a small paring knife, peel by rotating the fruit slowly in your hand; remove only the thin outside (colored) layer of skin. Scrape off any white membrane attached to the peel.

To dry at room temperature. Thread a large trussing needle with sewing thread, nylon thread or fishing line, or kitchen string. Push the needle through each strip of peel. Leave a space of about an inch between strips for air circulation. Hang garlands for 3 to 4 days in a warm, dry spot—an attic, covered porch, or kitchen that stays at 65° to 90°F—until strips are hard and completely dry. Store strips in an airtight container, or grate in a food processor or blender before storing if desired.

To dry in a dehydrator. Prepare strips as for drying at room temperature and set on trays in dehydrator for 2 to 6 hours at 120°F. Grated peel will dry in 1 to 4 hours in a 120°F dehydrator.

To dry in an oven. Place strips in a 120° to 140°F oven for 2 to 6 hours. Grated peel will dry in a 120° to 140°F oven in 1 to 4 hours.

To dry in a microwave oven. Spread grated peel in a single layer on paper toweling; top with a second sheet of toweling. Microwave on full power (high) for 3 to 4 minutes. If still slightly damp, finish drying peel at room temperature. Do not dry strips in a microwave.

HOW TO DRY CHILE PEPPERS

A string of chile peppers suspended from the kitchen ceiling is both decorative and functional. Whole chiles, mild or hot, dry in several weeks at room temperature. (Note: High humidity can cause peppers to spoil before they dry.) You can also dry whole, halved, or sliced chiles in a dehydrator, an oven, or the sun. California green and reddish Anaheim peppers and yellow, green, or orange jalapeño peppers dry well.

To dry at room temperature. Thread a trussing needle with kitchen string, nylon thread, or fishing line and push the needle through the stem of each chile. Position chiles alternately left and right. Tie a loop at the top of the string to hang chile garland. Hang chiles in a warm, dry area for about 3 weeks, until they shrivel and feel dry. Pick them off the string as needed; they will keep their flavor for at least a year. Use to flavor Mexican dishes, soups, stews, and sauces.

To dry in a dehydrator, an oven, or the sun. Whole, halved, or sliced chile peppers or whole peppers on a string will dry until brittle in 12 to 18 hours in an oven or a dehydrator set at 120°F. In a 140° oven, they'll dry in 2 to 5 hours. In the sun, chiles take 1 to 2 days to dry.

HOW TO DRY GARLIC AND SHALLOTS

Because garlic is readily available the year around, you will probably want to dry it only to make garlic powder or as garlic salt. For garlic powder, simply grind dried garlic in a blender or food processor; see the recipe for garlic salt on page 88. To dry garlic, peel garlic cloves. In a 120°F dehydrator, garlic will dry crisp in 6 to 8 hours. In a 140° oven, garlic will dry in 2 to 4 hours.

Dry fresh shallots following the instructions for drying garlic.

*Dry fiery fresh peppers—
and then crumble or grind
them for use in chili and
other spicy-hot dishes.*

ORNAMENTAL GOURDS

Ornamental gourds make beautiful autumn decorative arrangements, and they can be grown and dried easily at home. Gourds grow quickly from seed if they have their quota of heat, especially warm nights. Pick them when the stems turn brownish. Punch the end close to the stem with a long needle to let air inside. Then hang them, their stems tied with string, in a dry, well-ventilated place. The seeds will rattle when the gourds are fully dry.

GARLIC SALT

You can use your oven to dry garlic salt.

1 1/4 cups kosher salt
2/3 cup peeled garlic cloves

1. Pour salt into container of blender or food processor. Add garlic cloves and process to a paste.
2. Spread mixture on a parchment or waxed paper–lined baking sheet.
3. Dry in a 120° to 140°F oven for 1 to 3 hours, or until paste is completely dry. (It's not necessary to place a fan in front of the oven.)
4. Return dried paste to blender or food processor and grind it to a fine powder. Store in airtight container.

Yield: 1 1/4 cups

LEMON-SEASONED PEPPER

6 lemons
1 can (4 oz) ground black pepper
1/2 cup toasted sesame seed
1/4 cup celery seed
1/4 cup onion salt
1/4 cup salt
1 tablespoon garlic powder

1. Finely grate lemon peel. (You should have about 6 tablespoons grated peel.) On baking sheet, spread grated peel in a thin layer. Dry in 120° to 140°F oven for 1 to 4 hours. Or dry in a dehydrator at 120°F for 1 to 4 hours. Cool.
2. Combine dried peel and remaining ingredients, mixing thoroughly. Store tightly capped in small jars.

Yield: About 2 cups

DRYING FLOWERS

HOW TO MAKE POTPOURRI

Rose, violet, lavender, jasmine, and gardenia—heady scents of summer. These fragrances needn't fade with the coming of winter frosts if you know how to make potpourri. Its aromatic perfume can enhance your home the year around.

Any flower or herb, cultivated or wild, can be dried for its fragrance, color, or texture. A potpourri mixture is limited only by your imagination. And if the floral scents you prefer are not found in your area, you can purchase them at herb and craft stores.

There are three basic ingredients in every potpourri mixture: the *base flower*, such as rose or lavender, and complementary spices and herbs; the *aromatic oil*, such as rose oil or lilac oil; and the *fixative*, such as orrisroot, gum benzoin, sandalwood, or common salt. Oils are added to strengthen the scent of the base flower; however, they should be used judiciously. Too many different ones will be overpowering. The fixative blends the fragrances and retards evaporation of the oils. Fixatives and aromatic oils are available at herb shops and at some health food stores and drugstores.

Picking and drying flowers. Pick flowers in the morning after the dew has dried. Damp flowers will discolor or mold and ruin your potpourri. Pluck the petals from flowers that are in full bloom and dry them individually. Pansies, daisies, and flower buds may be dried whole.

To dry flowers, spread them out on paper towels in a cool, dry room, out of direct sunlight; they will dry in several days to a week. Or dry in a dehydrator at 120°F for 1 to 2 hours.

Mixing and blending the potpourri. Begin by mixing the dry ingredients—flower petals, herbs, spices, and so forth. Then mix the essential oils with the fixative in a glass or ceramic bowl. Add the oil-fixative mixture to the dry ingredients and mix thoroughly. Store the potpourri in a tightly closed jar for at least a week, stirring once or twice, to blend the scents.

Uses. After the fragrances have aged, the potpourri can be used in a variety of ways. Placed in decorative open containers, it acts as a subtle, natural room deodorizer. Glass bowls or pitchers of potpourri in the bathroom release their scent every time steam from the shower activates them. Potpourri tied in cheesecloth or muslin bags and placed in bath water sweetens it and refreshes the bather. Potpourri sachets scent lingerie and bedding. Your own handiwork with ribbon and lace can produce a nostalgic gift.

Try the potpourri recipes at right. These mixtures are just suggestions—the exact proportions are not important. Devise your own blends of flowers and aromatic oils.

HOW TO MAKE CRYSTALLIZED FLOWERS

Crystallized flowers make lovely edible decorations for elegant desserts. Choose white or brightly colored flowers with simple petal arrangements, such as small orchids, roses, sweet peas, or violets.

Applying the sugar coating. Place 1 egg white in a small bowl; stir lightly. Dip flowers, one at a time, in egg white or apply egg white with a small artist's brush; cover all parts of petals. Remove excess white that could cause petals to stick together.

Sprinkle or sift superfine sugar over petals. Cover all egg white, shaking to avoid clumping. Blow softly on flowers to remove excess sugar.

Drying the flowers. Place flowers on foil-lined baking sheet. Let dry in cool area for 2 to 3 days.

LEMON-ROSE BUD POTPOURRI

- 4 cups dried rose buds
- 1 cup lemon verbena
- 1 cup dried lavender flowers
   Finely grated peel of 1 orange
   Finely grated peel of 1 lemon
- 1 teaspoon ground allspice
- 1/2 teaspoon ground cloves
- 3 drops oil of lavender flowers
- 3 drops oil of roses
- 1 tablespoon orrisroot
- 1 tablespoon gum benzoin

1. Mix together rose buds, lemon verbena, lavender flowers, and orange and lemon peels. Let stand 2 days.

2. Stir in allspice and cloves.

3. Combine lavender and rose oils with orrisroot and gum benzoin in a small ceramic or glass bowl.

4. Add fixative-oil mixture to dry ingredients and mix thoroughly.

5. Store in a tightly closed jar 1 week or longer to blend scents. Shake jar or stir contents 4 or 5 times during storage.

Yield: About 6 cups

Use your imagination in blending dried flowers and aromatic oils to make heady potpourris.

SPICY ROSE POTPOURRI

- 2 cups dried rose petals
- 1/4 cup dried lavender flowers
- 1/4 cup lemon verbena
- 1/2 teaspoon ground cloves
- 1/2 teaspoon ground allspice
- 1 tablespoon grated dried orange peel (see page 86 for instructions)
- 2 teaspoons orrisroot
- 2 drops oil of lavender flowers
- 2 drops oil of roses

1. Mix together rose petals, lavender flowers, lemon verbena, cloves, allspice, and orange peel.

2. Combine orrisroot with lavender and rose oils in a small ceramic or glass bowl.

2. Add fixative-oil mixture to dry ingredients and mix thoroughly.

4. Store in a tightly closed jar 1 week or longer to blend scents. Shake jar or stir contents 4 or 5 times during storage.

Yield: About 2 1/2 cups

HISTORICAL NOTES

Centuries ago smoking (or, more accurately, smoke-curing) was discovered as a technique to preserve and flavor meat. In colonial America, affluent homes had their own smokehouses. And the chimneys of many cabins contained smoke holes where meat was hung, to tap the smoke and heat of the hearth.

When smokehouses became a thing of the past, enthusiastic sportsmen fabricated smokers from old refrigerators, barrels, or whatever for home-smoking, to preserve their catch and enjoy the inimitable flavor smoke lends to fish and game.

SMOKE-COOKING

Today, although smoking as a means of preserving foods is done mainly in commercial smokehouses, backyard chefs who love to barbecue are discovering smoke-cooking in special water smokers or covered barbecue grills. Charcoal (or in some cases gas or electricity) provides the heat to cook the food, while aromatic hardwood imparts a tantalizing aroma and flavor and liquid keeps the food moist.

EQUIPMENT

To smoke-cook, you'll need a water smoker (available at hardware, department, and gourmet cookware stores) or a covered barbecue grill. Most water smokers use charcoal as the heat source, but electric and gas models are available, too. Charcoal water smokers consist of a base pan, the smoker body, and a domed lid. The units are equipped with a fire pan, which fits inside the base pan to hold the charcoal, and a water pan and a cooking grill, which fit inside the smoker body to hold the liquid and the food. Large smoker models with more than one grill make it possible to smoke quantities of meat for a crowd or to cook both meat and vegetables at the same time.

COOKING FUELS

Pressed charcoal briquets and mesquite charcoal are the most commonly available cooking fuels. Mesquite charcoal, made from charred mesquite wood, is more expensive than pressed briquets, but many cooks prefer it because it imparts a non-smoky flavor all its own. Alder, hickory, and oak, available in chunks and as pressed logs, also are sold as barbecue fuel. Each of these woods contributes its own subtle flavor to the food as it provides the heat for cooking.

WOODS AND AROMATICS FOR FLAVORING

Any hardwood (from deciduous trees) is appropriate for smoking. Hickory is the most popular, but oak, pecan, walnut, alder, aspen, beech, and fruitwoods like apple, apricot, pear, or cherry can be used. You can also use grapevines. Wood for flavoring need not be completely dry; a few green twigs will enhance smoke production. Cut wood into approximately 2-inch chunks, small twigs, or chips or purchase packaged wood chunks, sticks, or chips where you find barbecue equipment. Softwoods and evergreens like pine, cedar, spruce, and fir are poor for smoking because these resinous woods give off a bitter-tasting pitch that taints the food and coats the inside of the smoker with a sticky black film.

Before adding them to the smoker, soak chunks or chips in water for 30 minutes so that they will smoke rather than burst into flames when placed on the cooking fuel. Don't oversoak—you'll waterlog the wood and disrupt the timing of the smoking process.

Start by adding four to six chunks or several handfuls (about 2 cups) of chips. As you become more experienced, you can use more or less flavoring wood to suit your taste.

Chunks burn slowly, so add all of them when you add the food. Small sticks and chips burn more quickly, so add some when you put the food on and the rest at intervals during cooking. (Be prepared to use up to 10 cups of chips during a 4- to 5-hour smoking period.) For an interesting touch, add four or five garlic cloves, four or five strips of lemon or orange peel, or three or four sprigs of rosemary or another woody herb to the coals along with the soaked wood.

HOW TO SMOKE-COOK IN A WATER SMOKER

In a water smoker, liquid is placed between the heat source and the food. As the liquid heats, it creates a moist, smoky atmosphere in which meat cooks tender and juicy without basting or turning. The

Grapevine chips, mesquite chunks, and hardwood logs combine with charcoal to create aromatic smoke that adds subtle flavors to foods.

A Water Smoker

Portable smoke-cookers—available at hardware, kitchenware, and department stores—use moisture to produce juicy smoke-flavored meats and vegetables. This cut-away illustration of a double-rack, charcoal-fueled model shows the standard parts.

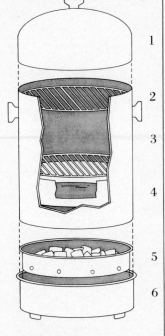

1 Domed lid
2 Cooking grill
3 Smoker body
4 Water pan
5 Fire pan
6 Base pan

liquid also keeps the smoking temperature low. The liquid can be water, fruit juice, wine, beer, marinade, or a combination of these.

To use a charcoal water smoker, light the cooking fuel and let it burn until the coals begin to gray. Then add the flavoring wood, the liquid, and the meat. Follow the manufacturer's guidelines as to how much charcoal, liquid, and flavoring wood to use. And resist the temptation to peek. Every time you lift the cover, you let heat, smoke, and moisture escape, and that slows cooking. Lift the cover only to add more liquid when necessary, generally after 3 or 4 hours of cooking.

How to Smoke-Cook in a Barbecue Grill

To convert a covered grill into a water smoker, build a fire at one end or on either side of a metal or foil drip pan. (This is the indirect cooking method outlined in manuals that come with covered barbecues.) Let the coals burn down until they are completely covered with gray ash and are low glowing. (If you cook the food over high heat, you are actually smoke-barbecuing—see explanation at right.) Fill the drip pan half full of liquid, position the grill about 6 inches above the pan, and place the meat directly over the pan. The liquid in the drip pan helps simulate the moist cooking environment of a water smoker, and the pan catches any meat drippings and basting sauces, eliminating flare-ups. Then cover the grill and adjust the dampers or vents so that they are open just a crack.

Foods cook more quickly in a covered barbecue than in a water smoker, and therefore take on a milder smoke flavor. To enhance the smokiness, you can add a few drops of natural liquid hickory smoke to the liquid in the drip pan, or to a marinade or basting sauce. Available at supermarkets, liquid smoke is actually condensed hickory wood smoke.

How to Smoke-Barbecue

You also can smoke-flavor foods simply by throwing a handful of wood chips directly on the hot coals or other heat source. This is called smoke-barbecuing or dry-smoking. You can smoke-barbecue on a barbecue grill—covered or not—or in a water smoker. Just eliminate the water pan and use the water smoker like a barbecue grill.

Any food that can be barbecued can be smoke-barbecued—from hamburgers to corn on the cob. Foods take on a much milder smoky flavor than when they are cooked in a water smoker.

Timing

Whether you use a water smoker or a barbecue grill, cooking times are approximate. A number of factors affect time, including the kind of fuel you use (heat intensity varies), the distance of the food from the coals, the size of the grill or smoker, the outdoor temperature, and the size and temperature of the meat to be cooked. Meat at room temperature cooks more evenly—so take small pieces from the refrigerator at least 15 minutes before cooking and remove large pieces an hour in advance.

The most dependable way to check meat for doneness is to use a meat thermometer. Keep in mind that the internal temperature of large pieces of meat cooked in a covered grill can rise 5° or more after they're removed from the grill. For perfect results with a barbecue grill, take the meat off just before it reaches the proper temperature. If you use a water smoker, however, wait until the meat reaches the proper temperature before you remove it.

Roasts, ribs, turkey, chicken, wild game, fish, and shellfish can be smoked to perfection; so can vegetables, and even nuts and cheese. The booklet that comes with your water smoker will give you instructions and times for all of these foods. The recipes on the opposite page are suited to either a covered barbecue grill or a water smoker.

HERB-SMOKED TURKEY

Allow ¾ to 1 pound of turkey per person. Leftovers make terrific sandwiches.

- *A 10- to 15-pound turkey (thawed if frozen), neck and giblets removed*
- *¹/2 cup butter or margarine*
- *¹/3 cup finely chopped mixed fresh herbs or ¹/2 teaspoon each dried thyme, sage, oregano, marjoram, and basil plus 2 tablespoons finely chopped fresh parsley*
- *2 cloves garlic, minced Salt and pepper*
- *2 cups white wine, at room temperature*

1. Soak 4 to 6 chunks of flavoring wood or 2 cups wood chips in water for 30 minutes. Drain.

2. Rinse turkey and pat dry. Tuck wing tips under back and tie drumsticks together.

3. Melt butter in a small saucepan. Add chopped herbs, garlic, and salt and pepper to taste.

4. *To smoke in a covered barbecue grill:* Build a charcoal fire on each side of a rectangular metal or foil drip pan, following manufacturer's directions for your barbecue. Pour wine into drip pan and add enough hot water to fill pan half full. When coals are covered with gray ash and are low glowing, add soaked wood (all the chunks; part of the chips). Oil grill and position it about 6 inches above drip pan. Brush turkey with some of the butter sauce; reserve remaining butter.

Place turkey, breast side up, on grill directly over drip pan. Cover grill and adjust dampers so that they are open just a crack. Cook turkey approximately 13 to 15 minutes per pound, basting 4 or 5 times with the butter, until joints move easily and juice runs clear when skin is pierced, or until a meat thermometer inserted in thigh registers 180°F. Add more briquets and wood chips after 45 to 60 minutes to maintain a constant low temperature.

To smoke in a water smoker: Follow manufacturer's directions for your smoker. Pour wine into water pan; add hot water to fill pan. Brush turkey liberally with butter sauce. Smoke-cook 6 to 8 hours, until a meat thermometer inserted in thigh registers 180°F. Add more liquid to pan after 3 or 4 hours of cooking if necessary.

5. Let turkey stand 10 to 20 minutes before carving.

Yield: 10 to 16 servings

TEXAS SMOKED BEEF RIBS

- *1 can (8 oz) tomato sauce*
- *2¹/2 tablespoons brown sugar*
- *1¹/2 to 2 tablespoons red wine vinegar*
- *1 tablespoon Dijon-style mustard*
- *2 cloves garlic, minced*
- *1 teaspoon Worcestershire sauce Salt and cayenne to taste*
- *4 to 6 pounds beef back rib bones, in slabs of 3 or 4 ribs each*

1. In a small saucepan over medium heat, combine tomato sauce, brown

sugar, vinegar, mustard, garlic, Worcestershire sauce, and salt and cayenne. Cool sauce slightly.

2. Place ribs in a baking dish. Spoon or brush sauce over ribs. Let ribs stand while you soak the flavoring wood and prepare the smoker or barbecue.

3. Soak 4 to 6 chunks of flavoring wood or 2 cups wood chips in water for 30 minutes. Drain.

4. *To smoke in a covered barbecue grill:* Build a charcoal fire on each side of a rectangular metal or foil drip pan, following manufacturer's directions for your barbecue. Pour water into drip pan until half full. When coals are covered with gray ash and are low glowing, add soaked wood (all the chunks; part of the chips). Oil grill and position it

If you plan to smoke-cook more than 4 pounds of these tangy barbecue-sauced ribs, you may need a rib rack to hold them.

about 6 inches above drip pan. Lift ribs from barbecue sauce and place on grill. Cover grill and adjust dampers so that they are open just a crack. Cook ribs 40 to 45 minutes for medium-rare, turning once or twice and brushing with any reserved sauce. Check ribs for doneness by cutting into meat. If using wood chips, add more once or twice during cooking.

To smoke in a water smoker: Follow manufacturer's directions for your smoker. Fill water pan with hot water and any sauce that remains after ribs are lifted from dish. Smoke-cook ribs 3 to 5 hours, or until meat pulls away from bones.

Yield: 4 to 6 servings

Index

Note: Page numbers
in italics refer to
illustrations.

U.S. Measure and Metric Measure Conversion Chart

Formulas for Exact Measures	Symbol	When you know:	Multiply by:	To find:		Rounded Measures for Quick Reference	
Mass *(Weight)*	oz	ounces	28.35	grams	1 oz	= 2 tbsp	= 30 g
	lb	pounds	0.45	kilograms	4 oz	= 1/2 c	= 115 g
	g	grams	0.035	ounces	8 oz	= 1 c	= 225 g
	kg	kilograms	2.2	pounds	16 oz	= 1 lb	= 450 g
					32 oz	= 2 lb	= 900 kg
					36 oz	= 2-1/4 lb	= 1000 g (1 kg)
Volume	tsp	teaspoons	5.0	milliliters	1/4 tsp	= 1/24 oz	= 1 ml
	tbsp	tablespoons	15.0	milliliters	1/2 tsp	= 1/12 oz	= 2 ml
	fl oz	fluid ounces	29.57	milliliters	1 tsp	= 1/6 oz	= 5 ml
	c	cups	0.24	liters	1 tbsp	= 1/2 oz	= 15 ml
	pt	pints	0.47	liters	1 c	= 8 oz	= 250 ml
	qt	quarts	0.95	liters	2 c (1 pt)	= 16 oz	= 500 ml
	gal	gallons	3.785	liters	4 c (1 qt)	= 32 oz	= 1 l
	ml	milliliters	0.034	fluid ounces	4 qt (1 gal)	= 128 oz	= 3-3/4 l
Length	in.	inches	2.54	centimeters	3/8 in.	= 1 cm	
	ft	feet	30.48	centimeters	1 in.	= 2.5 cm	
	yd	yards	0.9144	meters	2 in.	= 5 cm	
	mi	miles	1.609	kilometers	12 in (1 ft)	= 30 cm	
	km	kilometers	0.621	miles	1 yd	= 90 cm	
	m	meters	1.094	yards	100 ft	= 30 cm	
	cm	centimeters	0.39	inches	1 mi	= 1.6 km	
Temperature	°F	Fahrenheit	5/9 (after subtracting 32)	Celsius	32°F	= 0°C	
					68°F	= 20°C	
	°C	Celsius	9/5 (then add 32)	Fahrenheit	212°F	= 100°C	
Area	in.^{2}	square inches	6.452	square centimeters	1 in.^{2}	= 6.5 cm^2	
	ft^2	square feet	929.0	square centimeters	1 ft^2	= 930 cm^2	
	yd^2	square yards	8361.0	square centimeters	1 yd^2	= 8360 cm^2	
	a	acres	0.4047	hectares	1 a	= 4050 m^2	